Heat and Mass Transfer - From Fundamentals to Advanced Applications

Edited by Diana Enescu

Published in London, United Kingdom

Heat and Mass Transfer - From Fundamentals to Advanced Applications
http://dx.doi.org/10.5772/intechopen.1005936
Edited by Diana Enescu

Contributors
Abdelmalek Atia, Abderrahim Bakak, Christophe Josserand, Diana Enescu, Driss Lahboub, Fabián A. Ortega-Quintana, Hermawan, Ildikó-Renáta Száva, Ioan Száva, Jalalov Garib, Jatoth Heeraman, Julie Fichot, Leonardus Murialdo Fransislus Purwanto, Marie-Louise Saboungi, Oussama Zobiri, Rasulov Mahir, Rodolphe Heyd, Sandeep Chinta, Sinsoysal Bahaddin, Stephanus Evert Indrawan, Teofil-Florin Gălățanu

First published in London, United Kingdom, 2025 by IntechOpen
IntechOpen is the global imprint of INTECHOPEN LIMITED, registered in England and Wales, registration number: 11086078, 167-169 Great Portland Street, London, W1W 5PF, United Kingdom

For EU product safety concerns: IN TECH d.o.o., Prolaz Marije Krucifikse Kozulić 3, 51000 Rijeka, Croatia, info@intechopen.com or visit our website at intechopen.com.

British Library Cataloguing-in-Publication Data
A catalogue record for this book is available from the British Library

Heat and Mass Transfer - From Fundamentals to Advanced Applications
Edited by Diana Enescu
p. cm.
Print ISBN 978-1-83634-506-0
Online ISBN 978-1-83634-505-3
eBook (PDF) ISBN 978-1-83634-507-7

Meet the editor

Diana Enescu is an Associate Professor at Valahia University of Târgoviște, Romania. Her academic work focuses on thermal engineering, with expertise in thermoelectric systems, photovoltaic-thermal applications, and energy-efficient technologies. She has published extensively on these topics and serves as Editor and Guest Editor for international journals and scientific conferences. She is also currently working as a researcher at the National Institute of Metrological Research (INRIM) in Italy, contributing to various European projects related to advanced thermal energy systems. Her activity integrates teaching and applied research in thermal engineering. She has been included in the Stanford–Elsevier World Ranking of Top 2% Scientists based on scientific performance and citation metrics.

Contents

Preface

This edited volume, *Heat and Mass Transfer – From Fundamentals to Advanced Applications*, brings together a series of contributions that address different aspects of heat and mass transfer, combining theoretical modeling with practical applications. The volume is divided into two thematic sections, reflecting the complementary roles of numerical analysis and real-world implementations in solving engineering problems. The first section, *Numerical Methods and Analytical Approaches in Heat and Mass Transfer*, focuses on computational techniques, simulation strategies, and analytical methods for studying thermal phenomena. These chapters are aimed at researchers and engineers working on mathematical modeling, algorithm development, and simulation of thermal systems:

- *Thermal Conduction Fundamentals and Applications*

- *Numerical Methods and Computational Models with Python in Heat Transfer*

- *Numerical Modeling of Heat and Mass Transfer in the Processes of Oil Deposits Operation in a Deformable Environment*

- *Modern Dimensional Analysis Involved in Heat Transfer Problems*

The second section, *Engineering Applications of Heat and Mass Transfer*, presents case studies and applications in fields such as electronics, wearable technologies, building physics, and industrial drying. These chapters demonstrate how thermal knowledge is applied in practice to improve performance, energy efficiency, and process control:

- *Heat Transfer within MOSFET Nanotechnology*

- *Thermal Energy Conversion in Wearable Thermoelectric Generators*

- *Controlling Heat Transfer through Walls and Roofs to Regulate Indoor Temperature: Analysis Using HT-Flux Software*

- *Drying of Aqueous Films, an Application of Heat and Mass Transfer*

The chapters in this volume offer a clear and structured view of current developments in heat and mass transfer. Readers will find both simulation tools and practical strategies that can be applied across various disciplines, including materials and electronics, as well as civil engineering and process technology. I want to thank all contributing authors for their contributions and dedication. I would also like to express my appreciation to the editorial team at IntechOpen for their support and professionalism throughout the publication process.

Diana Enescu
Department of Electronics,
Telecommunications and Energy,
Valahia University of Târgoviște,
Târgoviște, Romania

Applied Metrology and Engineering Division,
National Metrology Institute,
Turin, Italy

Section 1

Numerical Methods and Analytical Approaches in Heat and Mass Transfer

Chapter 1

Thermal Conduction Fundamentals and Applications

Sandeep Chinta and Jatoth Heeraman

Abstract

Thermal conduction is a primary mode of heat transfer, occurring at the molecular level through the exchange of kinetic energy among particles through collisions and vibrations. Conduction is mostly seen in solids, although it also significantly occurs in fluids under certain circumstances. This chapter examines the basics of thermal conduction, emphasizing Fourier's law, which correlates the rate of heat transmission with material characteristics, temperature gradients, and cross-sectional area. Theoretical and experimental methods for forecasting and measuring thermal conductivity are examined, emphasizing their significance in engineering applications. Principal subjects include the function of conduction in both steady-state and transient circumstances, thermal resistance in multi-layered systems, and the influence of material characteristics on heat transfer efficacy. Practical applications are analyzed, including thermal management in electronic devices and the design of insulation and heat exchangers. Recent breakthroughs in materials, such as graphene and phase change composites, highlight their potential to improve thermal performance. This chapter emphasizes the need of enhancing conduction-based systems for greater energy efficiency and sustainability. Engineers may construct sophisticated systems for many purposes, such as renewable energy, electronics cooling, and building insulation, by comprehending thermal characteristics and using innovative materials. The ideas presented here advance thermodynamics and mechanical engineering, promoting novel solutions to current heat transfer issues.

Keywords: heat transfer, rate of conduction, conduction, heat transfer coefficient, thermal conductivity

1. Introduction

Conduction heat transfer is a fundamental mode of heat transfer where thermal energy is transferred through a material without any movement of the material itself. This transfer occurs due to the interaction of particles at the molecular or atomic level, where energy is passed through collisions and vibrations. Conduction is most significant in solids, as their tightly packed particles facilitate efficient energy transfer. Conduction heat transfer is a basic way that heat moves, mainly happening in solids because of how molecules vibrate and transfer energy to nearby particles. Even though it does not work as well in fluids and gases, it can still be important in certain situations. Conduction happens because of a temperature difference, making thermal energy move from hotter areas to cooler ones, in line with the second law of thermodynamics [1, 2].

The phenomenon is described by Fourier's law of heat conduction, which relates the heat transfer rate to the material's thermal conductivity, temperature gradient, and cross-sectional area. Materials with high thermal conductivity, such as metals, transfer heat more effectively, whereas insulating materials like wood or foam resist heat flow. Conduction plays a vital role in various engineering applications, including thermal insulation, heat exchanger design, and electronic cooling systems. Understanding conduction is crucial for optimizing energy efficiency and ensuring effective thermal management in mechanical and thermal systems. This makes it a key area of study in thermodynamics and mechanical engineering [3–5].

1.1 Heat transfer through conduction

Heat transfer through conduction is the process by which thermal energy is transferred between materials that are in direct contact or within a material. Energy exchange is facilitated by the kinetic interactions of particles, including elastic and inelastic collisions between atoms, as well as electron drift, in this mode of heat transfer that operates at the molecular level. The fundamental principle of conduction is that thermal energy is transferred from regions of higher temperature (higher energy) to regions of lower temperature (lower energy). Heat transfer does not occur when two regions are at the same temperature [6–8].

The vibrational energy of a material's molecules is directly proportional to its temperature. Energy is exchanged as these molecules interact, establishing a pathway for heat transfer. Joseph Fourier established Fourier's law, which mathematically describes this mechanism. Fourier's law offers a prognostic framework for estimating the rate of heat transfer through conduction [9, 10].

1.2 Fourier's law of heat conduction

The rate of heat transfer by conduction (Q) is proportional to the cross-sectional area (A) through which the heat flows and the temperature gradient (dT/dx) according to Fourier's law:

$$Q = -kA\frac{dT}{dx} \tag{1}$$

where:

- Q: Heat transfer rate (W),
- k: Thermal conductivity of the material (W/m K),
- A: Cross-sectional area perpendicular to the direction of heat transfer (m^2),
- dT/dx: Temperature gradient in the direction of heat transfer (K/m).

Thermal conductivity (k) is a material property that quantifies the capacity of a substance to conduct heat. Effective conductors are materials with high thermal conductivity, such as metals, while insulators are materials with low thermal conductivity, such as wood or foam [11]. The thermal conductivity unit is typically denoted as W/m. K in the SI system.

1.3 Heat conduction conditions

Solids, liquids, and gases undergo heat conduction when a material medium is present to facilitate the transfer. Nevertheless, conduction is not possible in a vacuum, as there are no particles to transfer energy. In gases and liquids, convective heat transfer frequently eclipses conduction, particularly when heat input induces fluid expansion. This expansion generates density gradients that stimulate fluid motion, thereby improving heat transfer through convection, which is generally more efficient than conduction in fluids [12, 13].

An example of heat conduction is the time-dependent temperature variation at a specific location within a material. This frequently occurs when a system undergoes a sudden change in its boundary conditions, such as a decrease in temperature or an increase in heat flow, for instance. A sharp spike in temperature serves as an illustrative example of this phenomenon. Analytical solutions for heat conduction problems are frequently restricted to straightforward geometries and boundary conditions. Numerical techniques such as finite element analysis and finite difference methods are frequently necessary for addressing increasingly complex scenarios. Comprehend the mechanisms by which electronic devices transfer heat during operation, evaluate the energy performance of structures, and analyze the thermal responses of materials to pulsed heat sources all of which require a grasp of heat conduction. The effective operation of numerous systems relies on accurate predictions of temperature distributions during heat conduction [14].

1.4 Prediction and measurement of thermal conductivity

The thermal conductivity of solids and liquids is often ascertained by experimental techniques since theoretical estimations frequently lack the requisite accuracy for practical use. Conversely, the thermal conductivity of gases at moderate temperatures strongly corresponds with the predictions of the kinetic theory of gases. This theory, which conceptualizes gas molecules as particles in perpetual random motion, offers a robust theoretical foundation for comprehending heat transport in gases. The kinetic hypothesis posits that the thermal conductivity of gases is mostly determined by molecular collisions and the transmission of kinetic energy among molecules. At moderate temperatures, when gas molecules display predictable behavior, the kinetic theory provides reliable predictions that facilitate more exact computations of thermal conductivity. This theoretical framework is especially efficient for gases under these circumstances, enabling more reliable calculations of their thermal characteristics than for solids and liquids, which often need empirical data [15, 16].

1.5 Practical applications

Heat conduction concepts are extensively applied across various engineering disciplines. Heat transfer mechanisms and methods to enhance their performance are essential for improving energy efficiency and sustainability [17–19].

1.5.1 Design of electronic thermal systems

Reliable thermal management is essential for the performance of electronic equipment. The demand for efficient cooling solutions is increasing due to the growing power density and miniaturization of electronic components. Thermal interface materials, heat sinks, and microchannel heat sinks are commonly employed to manage heat produced by electronic components. Attention to material properties, geometry, and fluid flow characteristics is crucial for the effective design of cooling systems. Electronic components are prevented from overheating and maintain reliable operation due to the high thermal conductivity of these materials, which facilitates effective heat transfer away from them. The design of electronic packaging significantly influences heat management. Optimal heat dissipation requires careful consideration of materials, component arrangement, and airflow pathways. Effective thermal management enhances the energy efficiency of electronic devices and concurrently boosts their reliability and lifespan [20, 21].

1.5.2 Building insulation

Heat conduction plays a critical role in building insulation, and insulation in buildings is engineered to reduce heat conduction from the interior to the exterior. This approach minimizes energy required for heating and cooling the building while ensuring a comfortable indoor temperature. Insulating materials characterized by low thermal conductivity, including fiberglass, cellulose, and polyurethane foam, effectively reduce heat transmission [22, 23].

1.5.3 Heat exchangers

The purpose of a heat exchanger is to facilitate the transfer of heat between two fluids [24]. Heat exchangers depend on thermal conduction for their design and operation. Heat exchangers are extensively utilized in various applications, such as power generation, refrigeration, and air conditioning. Optimizing surface area for heat transfer, minimizing thermal resistance, and selecting materials with high thermal conductivity are critical design considerations for heat exchangers [25]. Recent advancements in materials science have resulted in notable enhancements in the efficiency of heat exchangers. Recent advancements include the development of novel alloys and composites. Enhanced thermal conductivity, corrosion resistance, and other advantageous properties of these materials enable the development of more efficient and smaller heat exchangers [26].

Microchannel heat exchangers have garnered significant attention owing to their elevated heat transfer rates and high surface area-to-volume ratio. Common applications for these devices encompass electronic cooling and high-power laser systems. Such applications require instances characterized by very high heat transfer rates. Various industries utilize heat exchangers to enhance energy efficiency and promote sustainability. Heat exchangers must be designed and optimized as part of this procedure.

2. Literature review: Conduction heat transfer

The examination of conduction heat transfer includes a variety of material systems and engineering applications, with continuous research focused on improving thermal conductivity and maximizing heat transfer efficiency. This literature review presents a summary of significant findings from both recent and foundational works in the field, emphasizing advancements in materials, theoretical models, and applications [27].

2.1 Enhancement of thermal conductivity

Gong et al. [28] investigated the improvement of heat transfer efficiency in thermally conductive polymer composites through the reduction of interfacial thermal resistance. Their findings are consistent with the principles of conduction heat transfer and provide valuable insights for optimizing polymer composites in thermal applications. Dai et al. [29] illustrated the improvement of thermal conductivity in polymer composites *via* anisotropic graphene frameworks, marking a significant advancement for sophisticated thermal management systems. Tseng et al. [30] conducted a study on the integration of functionalized graphene oxide into polyimide nanocomposites, resulting in a notable improvement in thermal conductivity.

Xu [31] highlighted the significance of graphene-based materials in enhancing the thermal performance of thermal interface materials (TIMs) and epoxy composites, respectively, demonstrating their potential applications in contemporary electronics. Ming et al. [32] presented an innovative graphitization technique aimed at producing carbon fibers with exceptional thermal conductivity, thereby enhancing material performance in heat transfer applications. Jeong and Yu [33] conducted an analysis of the relationship between thermal conductivity and density in insulation materials, highlighting the significance of material properties in enhancing conduction heat transfer.

2.2 Thermal management in electronics

Li et al. [34] examined TIMs developed to improve conduction heat transfer in microprocessors, tackling significant challenges in thermal management through the integration of high thermal conductivity and elastic compliance. Zhang et al. [35] presented a multifunctional elastomer characterized by high thermal conductivity and electromagnetic wave absorption properties, emphasizing advancements in electronics cooling. Wingert et al. [36] conducted a comprehensive analysis of thermal transport in amorphous materials, emphasizing phonon interactions as a key mechanism for conductive heat transfer in semiconductors and similar materials.

2.3 Conduction in building and insulation materials

Tian et al. [37] conducted an analysis of the heat transfer coefficients associated with exterior wall insulation, highlighting the differences observed between theoretical predictions and empirical measurements. Their findings are essential for enhancing thermal performance in building materials. Jia and He [38] created a mathematical model to analyze heat transfer in refrigerated cargo holds on ships, highlighting the significance of choosing suitable insulating materials according to their thermal characteristics. Carson and Sekhon [39] presented precise models for assessing the thermal conductivity of particulate materials, delivering important insights for conduction heat transfer in multilayer insulation systems. Vėjelienė [40] and Abdou and Budaiwi [41] examined the influence of moisture content on the thermal conductivity of fibrous insulation materials, highlighting its significance in conduction and energy transfer within thermal applications.

2.4 Material design and theoretical models

Adepitan et al. [27] provided empirical data regarding the thermal conductivities of different insulating materials, highlighting the significance of experimental validation in conduction research. Mirabolghasemi et al. [42] emphasized that advanced architected cellular materials attain improved thermal conductivity *via* distinctive structural designs, consistent with the principles of conduction heat transfer. Ji and Huang [43] investigated graded materials for the manipulation of thermal conductivities, providing valuable insights into the control of conduction mechanisms for engineering applications. Mavromatidis et al. [44] conducted a comprehensive analysis of multilayer insulation, highlighting the interaction between conduction and radiation.

2.5 Emerging trends in thermal materials

Wang et al. [45] investigated advancements in high thermal conductivity insulation materials to tackle heat accumulation challenges in compact electronics, illustrating the significance of material innovations in thermal management. Yoshiizumi et al. [46] conducted an evaluation of the thermal properties of polymer thin films, highlighting the distinctions in conduction performance between thin films and bulk materials. Lebelo [47] investigated the thermal stability of reactive materials and cylindrical pipes, respectively, offering valuable insights into the interplay of conduction, convection, and radiation in complex systems. Kobari et al. [48] emphasized the significance of minimizing radiative heat transfer to improve insulation performance, thereby enriching the comprehension of conduction.

3. Thermal conductivity

3.1 Significance of thermal conductivity

Thermal conductivity is essential in the heat transmission mechanisms of conduction and convection. In standard situations, natural convection often governs heat transfer in fluids; but, in certain constrained conditions, such as airspaces with restricted airflow, conduction may emerge as the predominant mechanism of heat transfer [49]. This is particularly applicable in smaller, confined areas when the temperature gradient inside the fluid is sufficiently steep to render conduction more relevant than convection, a phenomenon that was first undervalued. Comprehending this transition is essential in the design of thermal systems, since dependence only on convection for heat disposal may result in erroneous assessments of thermal performance [50]. The thermal conductivity of materials, which dictates their capacity for heat transmission, must be meticulously evaluated in applications where heat transfer mostly happens by conduction [51]. This encompasses the design of heat exchangers, insulation systems, and other systems where thermal efficiency is critical. An accurate understanding of thermal conductivity and its significance in heat transfer is crucial for improving these systems to achieve efficient thermal management [52].

3.2 Mechanism of heat transfer

Thermal conduction takes place at the molecular level as kinetic energy is transferred from regions of higher energy (hotter) to regions of lower energy (colder). Energy transfer occurs *via* molecular collisions and vibrations, with a temperature differential being crucial for conduction to take place. Metals generally exhibit outstanding thermal conductivity as a result of their atomic structure. The materials possess free electrons that are not attached to any particular nucleus, enabling these electrons to move freely throughout the substance. During their movement, these free electrons convey energy through collisions with other electrons, ions, and nuclei within the metallic lattice. This phenomenon is similar to the mechanism of electrical conduction, which explains why many effective thermal conductors, like copper and aluminum, also serve as efficient electrical conductors.

3.3 Distinctive materials and their uses

Thermal conduction is a basic phenomenon in which heat energy is transferred through materials as a result of temperature gradients. Diverse materials have distinct heat conduction characteristics that make them essential for particular applications [53]. Diamond has remarkable heat conductivity, around five times that of copper, establishing it as one of the most effective thermal conductors. Moreover, its dielectric strength, almost tenfold that of rubber, renders it optimal for use in heat sinks for electronic equipment necessitating strong thermal conductivity in conjunction with electrical insulation. In addition to diamond, copper is a commonly used substance, celebrated for its superior heat conductivity [54]. Copper is widely used in heat exchangers, radiators, and electronic cooling systems. Nonetheless, its lack of electrical insulation restricts its use in devices that need both high heat conductivity and electrical isolation [55].

Silicon carbide (SiC) is distinguished by its superior thermal conductivity and remarkable resilience under harsh environments. These attributes render it essential in high-performance electronics, aerospace components, and power modules functioning at higher temperatures [56]. Novel materials such as graphene have remarkable possibilities in heat control. Graphene, known for its exceptional thermal conductivity and lightweight properties, is increasingly used in microelectronics, improved cooling systems, and high-speed computer devices where effective heat dissipation is essential [57]. This comprehensive discourse elucidates the varied characteristics of different materials, correlating them with heat conduction and their respective uses. This method emphasizes the significance of comprehending material-specific features to improve thermal management in various sectors [58].

3.4 Applications of thermal conductivity

Comprehending and enhancing thermal conductivity carry significant implications across various sectors. In electronics, materials exhibiting high thermal conductivity are utilized for the effective management of heat dissipation in compact and high-power devices. In building insulation, materials with low thermal conductivity are crucial for reducing energy loss. In addition, the aerospace and automotive sectors utilize lightweight materials with customized thermal properties to improve thermal efficiency while maintaining structural integrity. Engineers can design systems and select materials that optimize heat transfer performance by utilizing the principles of thermal conductivity, thereby enhancing energy efficiency and thermal management in various applications [59, 60].

Thermal conductivity is essential in conduction heat transfer and also affects convection in certain situations. Materials exhibiting high thermal conductivity, including metals and advanced composites, promote efficient heat transfer, while insulating materials characterized by low conductivity inhibit energy loss. The examination and implementation of thermal conductivity play a crucial role in the progression of energy systems and the improvement of thermal performance in engineering solutions [61, 62].

An important part of many engineering tasks is conducting electricity, especially in systems that control heat. As an example, conduction is used to move heat between fluids in heat exchanges, which makes energy systems more efficient in fields like HVAC and power generating. To keep heat from escaping, insulation materials in factories and houses also use convection to their advantage. This helps save energy. In electronics cooling, heat is removed through conduction to keep sensitive parts working reliably. This keeps electronics from burning and makes them last longer. The thermal conductivity of the materials used in these situations is very important because it tells us how well they can move heat away from parts that make heat. For example, copper and aluminum, which have a high thermal conductivity, are used in heat exchangers to make the heat transfer process as efficient as possible. These cases show how important conductivity is for making engineering systems work better and be more reliable. References that support these uses will be included so that you can get a better idea of how conduction is used in real-life engineering [63, 64].

This study investigates solutions to conduction problems using both analytical and numerical methods. Accurate modeling and simulation of heat transmission in many materials and systems is made possible by these technologies, such as computational fluid dynamics (CFD) and finite element analysis (FEA). Because of its ability to optimize design and anticipate thermal performance, computational fluid dynamics (CFD) has seen considerable use in the analysis of heat exchangers and thermal management systems. While numerical approaches are essential for handling complicated, real-world conduction issues, analytical methods, such as calculating the heat conduction equation, provide precise answers for basic geometries. As an example, the methods used to fix conduction problems in heat exchangers are related to the topics covered in Ref. [65], which is about improving heat exchanger performance using CFD and ANN. The use of thermal conductivity measurements is also shown in Ref. [66], which is crucial for verifying the heat transport models used in these approaches. Engineers may create systems that control heat transport and increase energy efficiency by integrating analytical and numerical methods.

4. Thermal resistances

Thermal resistance serves as an essential parameter for comprehending and measuring the resistance to heat transfer within a system. This method is especially effective in assessing the thermal performance of components, interfaces, and insulation materials in engineering applications. Thermal resistance can be understood through an analogy to electrical resistance, where the movement of heat parallels the flow of current, and the temperature gradient facilitates this movement.

Thermal resistance is typically represented in units of K/W, indicating the temperature increase per unit of heat transfer. The calculation can be performed using the following formula:

$$R_{th} = \frac{\Delta T}{Q} \tag{2}$$

In this context, R_{th} represents the thermal resistance, ΔT denotes the temperature difference across the material or interface, and Q signifies the heat transfer rate.

4.1 The significance of thermal resistance in heat transfer

Thermal resistance plays a vital role in applications that involve multi-layered materials or systems featuring multiple thermal interfaces. This defines the efficiency of heat transfer through these layers or across interfaces. A lower thermal resistance signifies enhanced heat transfer capability, whereas a higher thermal resistance denotes increased insulation or resistance to heat flow.

4.2 Categories of thermal resistance

Thermal resistance can be classified according to its application and the physical characteristics of the materials or systems involved.

Material thermal resistance: This characteristic is intrinsic to a material and is associated with its thermal conductivity "k." The thermal resistance for a homogeneous material with a thickness of L is expressed as follows:

$$R_{th} = \frac{L}{kA} \tag{3}$$

In this context, A represents the cross-sectional area through which heat flows.

Contact thermal resistance: This phenomenon takes place at the interface where two materials are in contact. Surfaces that may seem smooth often possess microscopic irregularities, which lead to air gaps and diminish heat transfer efficiency. Contact thermal resistance can be reduced by employing thermal interface materials (TIMs) such as thermal pads, greases, or adhesive bonding materials.

4.3 Applications of thermal resistance

Thermal resistance is essential in the design and optimization of thermal management systems. Primary applications encompass:

Electronic devices: In electronic components, including CPUs and power modules, the thermal resistance of heat sinks, thermal interface materials, and packaging materials plays a critical role in determining the efficiency of heat dissipation. Minimizing thermal resistance is essential to avoid overheating and guarantee the reliability of the device [67].

Insulation systems: In the context of building materials and refrigeration systems, thermal resistance is a critical factor that determines the insulating performance of materials, thereby impacting energy efficiency and temperature regulation [68].

Practical illustrations of electronic thermal design include:

Composite materials: Advanced composite materials frequently exhibit customized thermal resistance characteristics to optimize insulation and heat transfer needs in aerospace, automotive, and industrial sectors [69].

Adhesive bonding materials: Adhesives utilized in thermal management systems are required to possess low thermal resistance to ensure efficient heat transfer across bonded interfaces, while also providing mechanical stability [70].

4.4 Enhancing thermal resistance

To enhance thermal resistance, engineers concentrate on:

- *Material selection:* Selecting materials that possess suitable thermal conductivities to align with application specifications.
- *Interface design:* Reducing contact resistance through the application of thermal interface materials and achieving optimal surface finishes.
- *Geometry and thickness:* Modifying the dimensions and configurations of components to optimize heat transfer and mechanical performance.

Thermal resistance is a key principle in the analysis of heat transfer and the design of engineering systems. Understanding and managing thermal resistance allow engineers to improve the performance and efficiency of systems in a range of applications, including electronics cooling, energy-efficient buildings, and advanced material systems. The importance lies in guaranteeing reliability, safety, and sustainability within contemporary technological solutions [71].

5. Theoretical background

5.1 Thermal conductivity

It is a key property that characterizes a material's capacity to transfer heat in the presence of a temperature gradient under steady-state conditions. This property is essential in the application of Fourier's law for heat conduction. It, as a transport property, signifies the rate at which energy is transferred through diffusion.

In heat transfer analysis, the boundary conditions, including temperature and heat flux on the material's surface, play a crucial role in determining the system's behavior. For instance, in steady-state thermal conduction problems, the temperature distribution within a material is heavily influenced by the prescribed surface temperature. Similarly, in transient heat conduction and convection analyses, accurate boundary conditions are essential to predict the time-dependent temperature changes and heat transfer rates. These conditions are fundamental in various engineering applications such as thermal insulation, electronics cooling, and heat exchanger design.

Figure 1 illustrates an experimental setup designed to measure the thermal properties of a sample material under controlled conditions. The system comprises the following key components:

Sample holder (A): The central section houses the sample material under investigation, positioned between two conductive blocks to ensure uniform heat transfer.

Heater section (B): Positioned adjacent to the sample, this unit generates controlled heat for thermal conduction experiments.

Cooling section (C): Includes cooling mechanisms (e.g., fins or coolant flow) to maintain ambient temperature on the other side of the sample.

Temperature measurement probes: Thermocouples or similar sensors embedded at various points to record temperature differentials across the sample and adjacent sections.

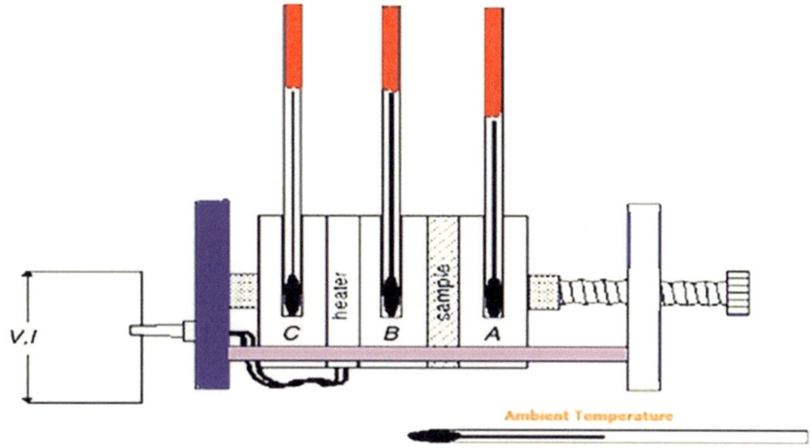

Figure 1.
Schematic diagram of the thermal conductivity measurement apparatus [20].

Power supply (V, I): Provides a regulated electrical current and voltage to the heater for precise thermal control.

Ambient temperature indicator: Indicates the baseline environmental temperature during the experiment for calibration purposes.

This setup is ideal for studying heat transfer properties, evaluating thermal conductivity, or performing material characterization tests. The precise control of input power and detailed temperature measurement ensure reliable and reproducible results.

In heat transfer analysis, the boundary conditions, including temperature and heat flux on the material's surface, play a crucial role in the study. The thermal boundary conditions are essential for accurately characterizing the material's thermal performance.

This graph illustrates, in **Figure 2**, the thermal conductivity values (in W/mK) of five different materials: Asbestos, P.O.P (Plaster of Paris), Hardboard, PVC wafer, and Paperboard. The x-axis represents the materials, while the y-axis indicates their

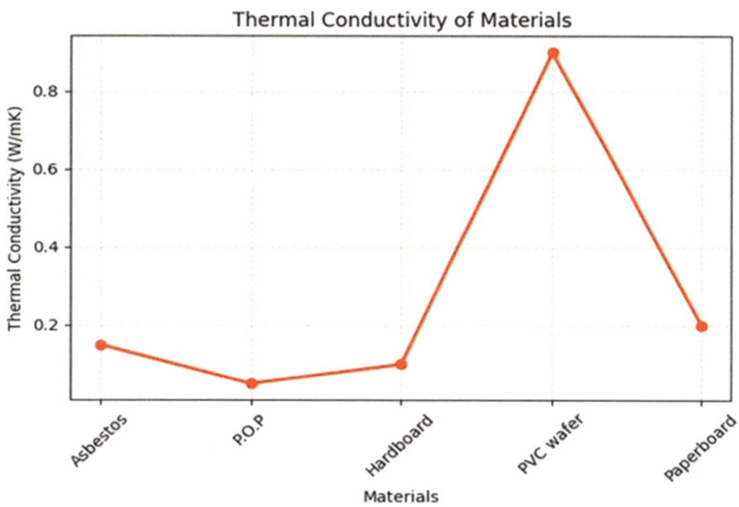

Figure 2.
Thermal conductivity of material.

thermal conductivity values. Among the materials, PVC wafer exhibits the highest thermal conductivity, significantly surpassing the others. In contrast, P.O.P shows the lowest thermal conductivity, indicating its better insulating properties. Asbestos, Hardboard, and Paperboard display moderate thermal conductivity values. The data suggests that PVC wafer is the least effective insulator, while P.O.P is the most effective.

The thermal conductivity of a substance is crucial for heat transmission and depends on factors such as microstructure, atomic structure, and the presence of impurities in the material. Crystalline materials exhibit superior thermal conductivity due to their high degree of organization, in contrast to amorphous materials. Grain boundaries and dislocations represent defects that can diminish heat conductivity through phonon scattering. Impurities can reduce heat conductivity by disrupting lattice vibrations. The thermal behavior of a material is influenced by factors beyond thermal conductivity, including density (ρ), specific heat capacity (c), and thermal diffusivity (α).

5.2 Thermal resistivity

Thermal resistivity is defined as the reciprocal of thermal conductivity, indicating a material's opposition to heat transfer. This can be articulated as:

$$r = \frac{1}{k} \tag{4}$$

where:

- r represents the thermal resistivity measured in m K/W.
- k denotes the thermal conductivity expressed in W/mK.

This graph presents, in **Figure 3**, the thermal resistivity values (in mK/W) of five materials: Asbestos, P.O.P (Plaster of Paris), Hardboard, PVC wafer, and Paperboard. The x-axis lists the materials, while the y-axis indicates their thermal resistivity values. Paperboard shows the highest thermal resistivity, indicating superior insulating performance. P.O.P follows closely with a high resistivity value. In contrast, PVC wafer has the lowest thermal resistivity, highlighting its poor insulating capabilities. Asbestos and Hardboard exhibit moderate resistivity values. This data suggests that Paperboard and P.O.P are the most effective insulators among the materials tested, while PVC wafer is the least effective.

5.3 Thermal diffusivity

Thermal diffusivity is a significant characteristic of insulating materials that assesses the rate at which heat spreads through a material. The ratio of heat produced to the heat stored in the material is represented as follows:

$$\propto = \frac{k}{\rho c_p} \tag{5}$$

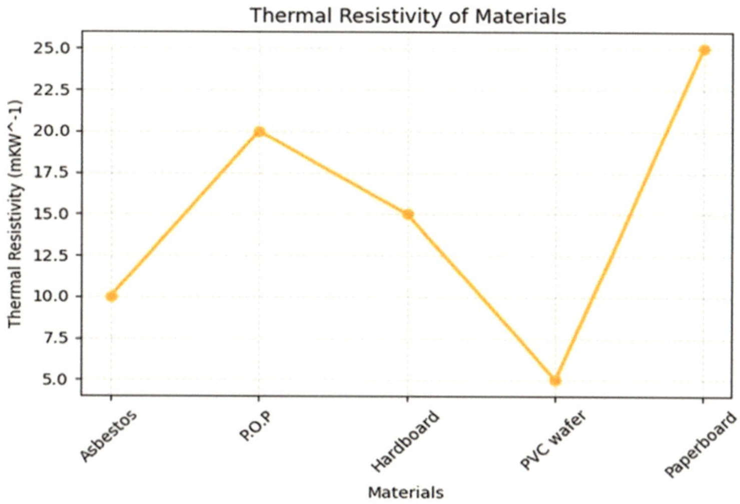

Figure 3.
Thermal resistivity of material.

where:

- k represents the thermal conductivity measured in W m^{-1} K^{-1},
- ρ = density of material in kg/m^3,
- c_p = Specific heat capacity at constant pressure (J/kg K),
- α = Thermal diffusivity (m^2/s).

The properties outlined are essential for the design and analysis of materials utilized in heat transfer applications, particularly in the context of insulating materials.

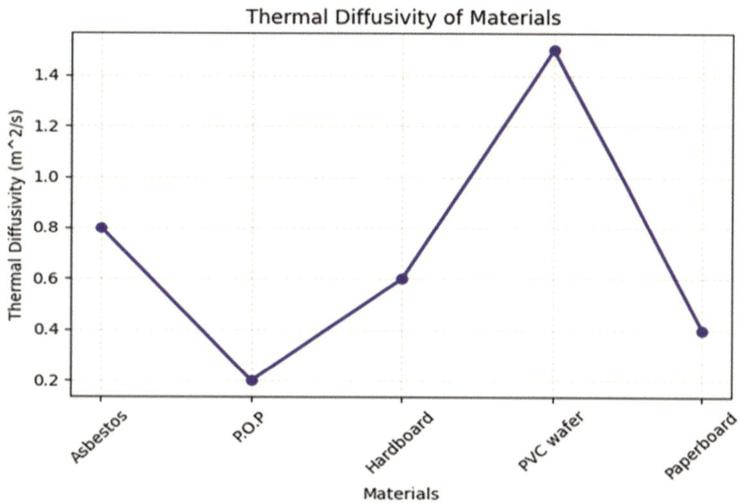

Figure 4.
Thermal diffusivity of material.

This graph displays, in **Figure 4**, the thermal diffusivity values (in m^2/s) of five materials: Asbestos, P.O.P (Plaster of Paris), Hardboard, PVC wafer, and Paperboard. The x-axis represents the materials, while the y-axis indicates their thermal diffusivity values. PVC wafer demonstrates the highest thermal diffusivity, suggesting its ability to quickly respond to temperature changes. In contrast, P.O.P exhibits the lowest thermal diffusivity, indicating slower heat propagation. Asbestos, Hardboard, and Paperboard show moderate thermal diffusivity values, with Asbestos being relatively higher than the others. These results highlight PVC wafer as the most responsive material to thermal variations, while P.O.P is the least responsive.

5.4 Density

Density is an important attribute of insulating materials and is essential for comprehending their heat transmission properties and is represented in **Figure 5**. It is defined as the mass of a substance per unit volume and is mathematically expressed as presented in Eq. (6).

$$\rho = \frac{m}{V} \tag{6}$$

where:

- ρ represents the density of the substance in kg/m^3,
- where m represents the mass of the substance in kg,
- V denotes the volume of the material in m^3.

In the domain of thermal insulation, density affects the efficacy of insulating materials. Materials with higher density often possess superior thermal conductivity, facilitating more efficient heat transmission, while lower-density materials usually have enhanced insulating characteristics due to their capacity to retain air pockets.

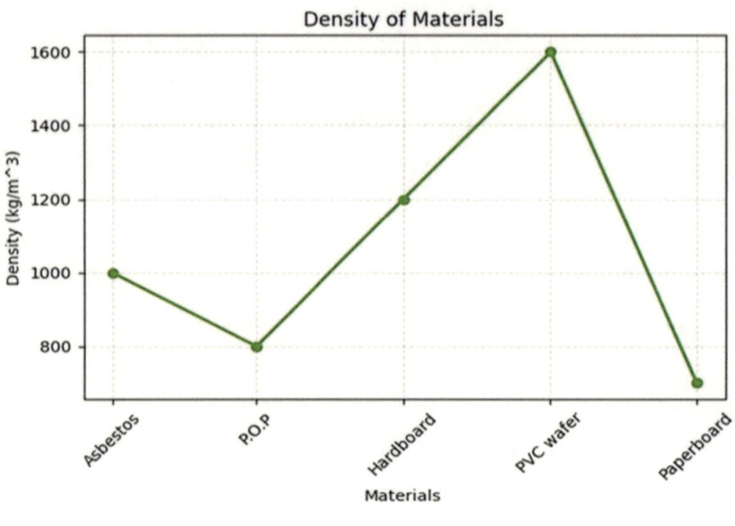

Figure 5.
Density of material.

Material	Thermal conductivity (k) ($W\,m^{-1}K^{-1}$)	Thermal resistivity (R) ($m\,K\,W^{-1}$)	Thermal diffusivity (α) ($\times 10^{-7}\,m^2/s$)
Asbestos	0.1717	5.8244	0.1352
P.O.P	0.1733	5.7709	1.2332
Hardboard	0.4197	2.3828	2.6861
PVC wafer	1.6499	0.6061	7.1883
Paperboard	0.2047	4.8852	1.9083

Table 1.
Computed thermal properties of insulating materials.

Density is directly correlated with the material's thermal diffusivity (α), as seen in Eq. (5), whereby the rate of heat conduction is influenced by the density of a material through the material.

The density of an insulating substance dictates the energy storage capacity per unit volume of the material. Low-density, high-porosity materials, such as foam-based insulators, efficiently diminish heat transmission and are often used in applications necessitating substantial thermal resistance.

The significance of density in insulating materials is attributed to its interplay with other parameters, including thermal conductivity and specific heat capacity. An accurate assessment of density guarantees that the insulating material delivers maximum thermal efficiency for a specific application, including building envelopes, refrigeration systems, or electrical equipment.

Table 1 displays the thermal properties of various insulating materials, including thermal conductivity (k), thermal resistivity (R), and thermal diffusivity (α). The values presented in the table were obtained through standardized experimental methods:

Thermal conductivity (k): Measured using the steady-state method, typically through a guarded hot plate or heat flow meter, to determine how well the material conducts heat.

Thermal resistivity (R): Calculated as the reciprocal of thermal conductivity ($R = 1/k$), representing the material's resistance to heat flow.

Thermal diffusivity (α): Determined using transient techniques, such as the laser flash method, which measures the speed at which heat spreads through the material.

6. Conclusion

This chapter provides a comprehensive overview of the concepts and practical applications of heat conduction. Various methods for measuring and predicting thermal conductivity have been investigated, including an analysis of Fourier's law, which governs heat transfer in conductive materials. The significance of understanding steady-state and heat conduction conditions, as well as the concept of heat resistance in multi-layered systems, has been highlighted. Material properties such as specific heat capacity, thermal diffusivity, density, and thermal conductivity have been analyzed for their impact on heat transfer efficiency. Real-world applications of thermal conduction include heat exchangers, electronic thermal design, and building insulation.

This chapter emphasizes the enhancement of conduction-based systems to achieve improved energy efficiency and overall sustainability. Through mastery of fundamental principles of thermal conduction and application of advanced materials and design strategies, engineers can develop intricate thermal management systems that tackle contemporary challenges and contribute to more sustainable future. The continuous advancement in material science and design methodologies is essential for enhancing thermal systems across various applications, such as renewable energy production, electronic cooling, and building insulation. Incorporating these advancements is essential for achieving substantial improvements in energy efficiency and reducing the environmental impact of diverse technologies.

Acknowledgements

We would like to express our sincere gratitude to the following individuals and institutions for their invaluable support during the development of this book chapter. Special thanks to our mentors and colleagues at the Department of Mechanical Engineering, Institute of Aeronautical Engineering, Hyderabad, India, whose guidance and insightful feedback were essential throughout the process. We also acknowledge the resources and support provided by Lovely Professional University, Phagwara Punjab, India, which enabled the research and writing of this chapter. Finally, we are deeply grateful to our families and friends for their constant encouragement and understanding, which made this contribution to "Heat and Mass Transfer—From Fundamentals to Advanced Applications."

Conflict of interest

The authors declare that there is no conflict of interest regarding the publication of this chapter. All data, analysis, and conclusions presented in this work are solely the result of the authors' research and are free from any commercial or financial relationships that could be construed as a potential conflict.

Notes/thanks/other declarations

We express our heartfelt gratitude to all individuals and organizations who contributed to the successful completion of this book chapter. Our deepest thanks go to our institutions for their unwavering support and for providing the necessary facilities that enabled this research.

We are especially thankful to the editorial team of Heat and Mass Transfer—From Fundamentals to Advanced Applications for their expert guidance and meticulous review process, which have significantly improved the quality of this chapter.

Lastly, we extend our appreciation to our colleagues, peers, and families for their encouragement and support throughout this endeavor. Their contributions have been invaluable in bringing this work to fruition.

Use of artificial intelligence

The authors confirm that they did not use artificial intelligence technologies when creating the current work.

DOI: http://dx.doi.org/10.5772/intechopen.1009788

Author details

Sandeep Chinta[1*] and Jatoth Heeraman[2]

1 Department of Mechanical Engineering, Institute of Aeronautical Engineering, Hyderabad, India

2 Department of Mechanical Engineering, Lovely Professional University, Phagwara, Punjab, India

*Address all correspondence to: chintasunny2020@gmail.com

References

[1] Kämpf H, Karsten G. Effects of different types of void volumes on the radial temperature distribution of fuel pins. Nuclear Applications and Technology. 1970;**9**(3):288-300. DOI: 10.13182/NT70-A28783

[2] Todreas NE, Kazimi MS. Nuclear Systems Volume I: Thermal Hydraulic Fundamentals. Taylor and Francis Group, CRC Press; 2021. DOI: 10.1201/9781351030502

[3] Majumdar A, Tien CL. Fractal network model for contact conductance. Journal of Heat Transfer. 1991;**113**: 516-525. DOI: 10.1115/1.2910594

[4] Haji-Sheikh A, Sparrow EM. The solution of heat conduction problems by probability methods. Journal of Heat Transfer. 1967:121-130. DOI: 10.1115/1.3614330

[5] Zohuri B, McDaniel P. Thermodynamics in Nuclear Power Plant Systems. New York, NY: Springer; 2015

[6] Isachenko VP, Osipova VA, Sukomel AS. Heat Transfer. Nirali Prakashan; 1980. Available from: books.google.com

[7] Zohuri B. Heat Pipe Design and Technology. FL: Taylor and Francis Group, LLC; 2011

[8] Sundén B. Introduction to Heat Transfer. Wit Press; 2012. Available from: books.google.com

[9] Bheda B, Tronstad J, Emady H. An experimental setup for studying conduction and radiation heat transfer in a rotary drum. Chemical Engineering Science. 2025;**302**:120786. DOI: 10.1016/j.ces.2024.120786

[10] Wang Y-B, Liu H-X, Wang C-X, Yan K-C, Wang S-Y, Zhang B-X, et al. Effect of electrode arrangement on heat transfer enhancement by electrohydrodynamic conduction pumps in a rectangular channel. Applied Thermal Engineering. 2025;**259**:124884. DOI: 10.1016/j.applthermaleng.2024.124884

[11] Hakvoort G, Van Reijen LL, Aartsen AJ. Measurement of the thermal conductivity of solid substances by DSC. Thermochimica Acta. 1985;**93**:317-320. DOI: 10.1016/0040-6031(85)85081-4

[12] Hou Y, Zhang X, Wang S. A stabilized state-based peridynamic heat conduction model for interface thermal resistance problems. Applied Mathematical Modelling. 2025;**137**:115504. DOI: 10.1016/j.apm.2024.05.001

[13] Orynyak I, Tsybulnyk A, Danylenko K, Oryniak A, Radchenko S. Timestep-dependent element interpolation functions in the method of matched sections on the example of heat conduction problem. Journal of Computational and Applied Mathematics. 2025;**456**:116222. DOI: 10.1016/j.cam.2024.116222

[14] Gao X-W, Huang S-Z, Cui M, Ruan B, Zhu Q-H, Yang K, et al. Element differential method for solving general heat conduction problems. International Journal of Heat and Mass Transfer. 2017;**115**:882-894

[15] Chiu J, Fair PG. Determination of thermal conductivity by differential scanning calorimetry. Thermochimica Acta. 1979;**34**(2):267-273. DOI: 10.1016/0040-6031(79)87116-6

[16] Bioucas B, Francisco E, Rausch MH, Schmidt J, Bück A, Koller TM, et al.

Effective thermal conductivity of nanofluids: Measurement and prediction. International Journal of Thermophysics. 2020;**41**:1-27. DOI: 10.1007/s10765-020-2621-2

[17] Klinar K, Swoboda T, Rojo MM, Kitanovski A. Fluidic and mechanical thermal control devices. Advanced Electronic Materials. 2020;**7**(3). DOI: 10.1002/aelm.202000623

[18] Haeger T, Heiderhoff R, Riedl T. Thermal properties of metal-halide perovskites. Journal of Materials Chemistry C. 2020;**8**:14289-14311. DOI: 10.1039/d0tc03754k

[19] Mashekova A, Zhao Y, Ng EYK, Zarikas V, Fok SC, Mukhmetov O. Early detection of the breast cancer using infrared technology—A comprehensive review. Thermal Science and Engineering Progress. 2021;**27**:101142. DOI: 10.1016/j.tsep.2021.101142

[20] Ellison GN. Thermal Computations for Electronics: Conductive, Radiative, and Convective Air Cooling. Taylor and Francis Group, CRC Press; 2020. DOI: 10.1201/9781003029328

[21] Price DM, Jarratt M. Thermal conductivity of PTFE and PTFE composites. Thermochimica Acta. 2002;**392**:231-236. DOI: 10.1016/S0040-6031(02)00105-3

[22] Vitiello D, NatAli B, TessierDoyen N, Tonnesen T, Lam L, Rebouillat L, et al. Thermal conductivity of insulating refractory materials: Comparison of steady-state and transient measurement methods. Open Ceramics. 2021;**6**:100118. DOI: 10.1016/j.oceram.2021.100118

[23] Zhou Y et al. Recent advances in thermal interface materials. ES Materials & Manufacturing. 2020;**7**(7):4-24. DOI: 10.30919/esmm5f717

[24] Zhang Y, Shi W, Zhang Y. From heat exchanger to heat adaptor: Concept, analysis and application. Applied Energy. 2014;**115**:272-279

[25] Harris M, Wu H, Zhang W, Angelopoulou A. Overview of recent trends in microchannels for heat transfer and thermal management applications. Chemical Engineering and Processing - Process Intensification. 2022;**181**:109155. DOI: 10.1016/j.cep.2022.109155

[26] Mangrulkar CK, Dhoble AS, Chamoli S, Gupta A, Gawande VB. Recent advancement in heat transfer and fluid flow characteristics in cross flow heat exchangers. Renewable and Sustainable Energy Reviews. 2019;**113**:109220

[27] Adepitan JO, Ogunsanwo FO, Ayanda JD, Okusanya AA, Adelaja AD, Oni OO, et al. Determination of thermal properties of some ceiling material commonly used in Ijebu-Ode, Nigeria. Nigeria Journal of Pure and Applied Physics. 2019;**9**(1):23-27. DOI: 10.4314/njpap.v9i1.5

[28] Gong P, Li L, Fu G-e, Shu S, Li M, Wang Y, et al. Highly flexible cellulose nanofiber/single-crystal nanodiamond flake heat spreader films for heat dissipation. Journal of Materials Chemistry C. 2022;**10**(33):12070-12079. DOI: 10.1039/D2TC01830F

[29] Dai W, Lv L, Ma T, Wang X, Ying J, Yan Q, et al. Multiscale structural modulation of anisotropic graphene framework for polymer composites achieving highly efficient thermal energy management. Advanced Science. 2021;**8**(7):2003734. DOI: 10.1002/advs.202003734

[30] Tseng I-H, Chang J-C, Huang S-L, Tsai M-H. Enhanced thermal conductivity and dimensional stability

of flexible polyimide nanocomposite film by addition of functionalized graphene oxide. Polymer International. 2013;**62**(5):827-835. DOI: 10.1002/pi.4375

[31] Xu K, Zhang Z, Wang Y, Li M, Chen Y, Kong X, et al. Sandwich-structured thermal Interface materials with high thermal conductivity. ACS Applied Engineering Materials. 2024;**2**(6):1572-1581. DOI: 10.1021/acsaenm.4c00124

[32] Ming X, Wei A, Liu Y, Peng L, Li P, Wang J, et al. 2D-topology-seeded graphitization for highly thermally conductive carbon fibers. Advanced Materials. 2022;**34**(28):2201867. DOI: 10.1002/adma.202201867

[33] Jeong YS, Ki Hyung Y. Experimental study of thermal conductivity of insulation materials made of expanded polypropylene, ethylene-vinyl acetate co-polymer and polyethylene. Advanced Materials Research. 2014;**831**:40-43. DOI: 10.4028/www.scientific.net/AMR.831.40

[34] Li J, Zhang Y, Liang T, Bai X, Pang Y, Zeng X, et al. Thermal interface materials with both high through-plane thermal conductivity and excellent elastic compliance. Chemistry of Materials. 2021;**33**(22):8926-8937. DOI: 10.1021/acs.chemmater.1c03275

[35] Zhang Z, Wang J, Shang J, Xu Y, Wan Y-J, Lin Z, et al. A through-thickness arrayed carbon fibers elastomer with horizontal segregated magnetic network for highly efficient thermal management and electromagnetic wave absorption. Small. 2023;**19**(4):2205716. DOI: 10.1002/smll.202205716

[36] Wingert MC, Zheng J, Kwon S, Chen R. Thermal transport in amorphous materials: A review. Semiconductor Science and Technology. 2016;**31**(11):113003. DOI: 10.1088/0268-1242/31/11/113003

[37] Tian YL, Lu JM, Zhang WY. Difference between the measured and theoretical value and theirs influencing factors for heat transfer coefficient of exterior wall external insulation. Advanced Materials Research. 2012;**415**:1427-1430. DOI: 10.4028/www.scientific.net/AMR.415-417.1427

[38] Jia JF, He W. Simulation study on heat insulating material used in refrigerated cargo hold shipboard of fishing vessel. Advanced Materials Research. 2011;**311**:1953-1956. DOI: 10.4028/www.scientific.net/AMR.311-313.1953

[39] Carson JK, Sekhon JP. Simple determination of the thermal conductivity of the solid phase of particulate materials. International Communications in Heat and Mass Transfer. 2010;**37**(9):1226-1229. DOI: 10.1016/j.icheatmasstransfer.2010.07.024

[40] Vėjelienė J. Processed straw as effective thermal insulation for building envelope constructions. Engineering Structures and Technologies. 2012;**4**(3):96-103. DOI: 10.3846/2029882X.2012.730286

[41] Abdou A, Budaiwi I. The variation of thermal conductivity of fibrous insulation materials under different levels of moisture content. Construction and Building Materials. 2013;**43**:533-544. DOI: 10.1016/j.conbuildmat.2013.02.058

[42] Mirabolghasemi A, Akbarzadeh H, Rodrigue D, Therriault D. Thermal conductivity of advanced architected cellular materials. In: Proceedings of The Canadian Society for Mechanical Engineering International Congress 2018

CSME International Congress 2018 May 27-30, 2018, Toronto, ON, Canada. 2018

[43] Ji Q, Huang J-P. Controlling thermal conduction by graded materials. Communications in Theoretical Physics. 2018;**69**(4):434. DOI: 10.1088/0253-6102/69/4/434

[44] Mavromatidis LE, Michel P, El Mankibi M, Santamouris M. Study on transient heat transfer through multilayer thermal insulation: Numerical analysis and experimental investigation. In: Building Simulation. Vol. 3. Springer, Tsinghua Press; 2010. pp. 279-294

[45] Wang Y, Tao ZQ, Chen JS. High thermal conductive insulation composite materials. In: Materials Science Forum. Vol. 848. Publisher in Material Science and Engineering, Trans Tech Publications Ltd; 2016. pp. 64-71. DOI: 10.4028/www.scientific.net/MSF.848.64

[46] Yoshiizumi M, Oikawa N, Shimada K, Endo S, Sugiyama M, Mikami H, et al. Thermal conductivity evaluation of polymer thin film. Transactions of the Materials Research Society of Japan. 2013;**38**(4):555-559. DOI: 10.14723/tmrsj.38.555

[47] Lebelo RS. Thermal conductivity impact on thermal stability of reactive materials. Diffusion Foundations. 2017;**11**:1-10. DOI: 10.4028/www.scientific.net/DF.11.1

[48] Kobari T, Okajima J, Komiya A, Maruyama S. Evaluation of radiative heat transfer in high-temperature porous insulation materials by using diffusion approximation. Japan Journal of Thermophysical Properties. 2015;**28**(4):179-184

[49] Lin Y, Jia Y, Alva G, Fang G. Review on thermal conductivity enhancement, thermal properties and applications of phase change materials in thermal energy storage. Renewable and Sustainable Energy Reviews. 2018;**82**:2730-2742. DOI: 10.1016/j.rser.2017.10.002

[50] Ebadi-Dehaghani H, Nazempour M. Thermal conductivity of nanoparticles filled polymers. In: Smart Nanoparticles Technology. London, UK: IntechOpen; 2012

[51] Asadi I, Shafigh P, Hassan ZFBA, Mahyuddin NB. Thermal conductivity of concrete—A review. Journal of Building Engineering. 2018;**20**:81-93. DOI: 10.1016/j.jobe.2018.07.002

[52] Chekhonin E, Popov Y, Peshkov G, Spasennykh M, Popov E, Romushkevich R. On the importance of rock thermal conductivity and heat flow density in basin and petroleum system modelling. Basin Research. 2020;**32**(5):1261-1276. DOI: 10.1111/bre.12427

[53] Kaviany M. Essentials of Heat Transfer: Principles, Materials, and Applications. Cambridge University Press; 2011. Available from: books.google.com

[54] Yoshida K, Morigami H. Thermal properties of diamond/copper composite material. Microelectronics Reliability. 2004;**44**(2):303-308. DOI: 10.1016/S0026-2714(03)00215-4

[55] Tian J, Kim T, Lu TJ, Hodson HP, Queheillalt DT, Sypeck DJ, et al. The effects of topology upon fluid-flow and heat-transfer within cellular copper structures. International Journal of Heat and Mass Transfer. 2004;**47**(14-16):3171-3186. DOI: 10.1016/j.ijheatmasstransfer.2004.02.010

[56] Singh D, Timofeeva E, Yu W, Jules Routbort D, France DS, Lopez-Cepero JM. An investigation

of silicon carbide-water nanofluid for heat transfer applications. Journal of Applied Physics. 2009;**105**(6). DOI: 10.1063/1.3082094

[57] Ghosh S, Nika DL, Pokatilov EP, Balandin AA. Heat conduction in graphene: Experimental study and theoretical interpretation. New Journal of Physics. 2009;**11**(9):095012. DOI: 10.1088/1367-2630/11/9/095012

[58] Mallik S, Ekere N, Best C, Bhatti R. Investigation of thermal management materials for automotive electronic control units. Applied Thermal Engineering. 2011;**31**(2-3): 355-362. DOI: 10.1016/j.applthermaleng.2010.09.023

[59] Zheng Q, Hao M, Miao R, Schaadt J, Dames C. Advances in thermal conductivity for energy applications: A review. Progress in Energy. 2021;**3**(1):012002. DOI: 10.1088/2516-1083/abd082

[60] Burger N, Laachachi A, Ferriol M, Lutz M, Toniazzo V, Ruch D. Review of thermal conductivity in composites: Mechanisms, parameters and theory. Progress in Polymer Science. 2016;**61**:1-28. DOI: 10.1016/j.progpolymsci.2016.05.001

[61] Younes H, Mingyang Mao SM, Murshed S, Lou D, Hong H, Peterson GP. Nanofluids: Key parameters to enhance thermal conductivity and its applications. Applied Thermal Engineering. 2022;**207**:118202. DOI: 10.1016/j.applthermaleng.2022.118202

[62] Cernuschi F, Ahmaniemi S, Vuoristo P, Mäntylä T. Modelling of thermal conductivity of porous materials: Application to thick thermal barrier coatings. Journal of the European Ceramic Society. 2004;**24**(9):2657-2667. DOI: 10.1016/j.jeurceramsoc.2003.09.012

[63] Hussain AR, Jahir AA, Alahyari SA, Eastman CT-E, Johnston S, Sobkowicz MJ. Review of polymers for heat exchanger applications: Factors concerning thermal conductivity. Applied Thermal Engineering. 2017;**113**:1118-1127. DOI: 10.1016/j.applthermaleng.2016.11.041

[64] Yang Y. Thermal conductivity. In: Physical Properties of Polymers Handbook. New York, NY: Springer New York; 2007. pp. 155-163

[65] Heeraman J, Kumar R, Chaurasiya PK, Verma TN, Chauhan DK. Optimisation and comparison of performance parameters of a double pipe heat exchanger with dimpled twisted tapes using CFD and ANN. Proceedings of the Institution of Mechanical Engineers, Part E: Journal of Process Mechanical Engineering. 2024;**238**:09544089231223599. DOI: 10.1177/09544089231223599

[66] Brennan WP, Miller B, Whitwell JC. Thermal conductivity measurements with the differential scanning calorimeter. Journal of Applied Polymer Science. 1968;**12**(7):1800-1802. DOI: 10.1002/app.1968.070120731

[67] Ocak M. Conduction based compact thermal modeling for thermal analysis of electronic components [master's thesis]. ProQuest Dissertations & Theses. Middle East Technical University; 2010. p. 31669545

[68] Lee SW, Lim CH. Reflective thermal insulation systems in building: A review on radiant barrier and reflective insulation. Renewable and Sustainable Energy Reviews. 2016;**65**:643-661. DOI: 10.1016/j.rser.2016.07.002

[69] Pietrak K, Wisniewski TS. Methods for experimental determination of solid-solid interfacial thermal resistance

with application to composite materials. Journal of Power Technologies. 2014;**94**(4):270

[70] Rabilloud G. Heat-resistant adhesives. In: Journal of Handbook of Adhesives and Sealants. Vol. 2. Elsevier Science Ltd; 2006. pp. 233-302. DOI: 10.1016/S1874-5695(06)80015-1

[71] Godi NY. Thermal internal resistance minimisation in microchannel heat sink with perforated micro fins geometry. International Journal of Heat and Fluid Flow. 2024;**108**:109487. DOI: 10.1016/j.ijheatfluidflow.2024.109487

Chapter 2

Numerical Methods and Computational Models with Python in Heat Transfer

Fabián A. Ortega-Quintana

Abstract

This chapter describes the most used numerical methods for solving engineering problems involving heat transfer mechanisms (conduction, convection, and radiation). It also presents simple computational models for solving these problems using Python, a language increasingly popular in industry, research, and education. The chapter covers problems involving nonlinear algebraic equations, systems of nonlinear algebraic equations, ordinary differential equations, and systems of ordinary differential equations of dynamic systems, as well as partial differential equations in both dynamic and static systems. Finally, it presents a specific methodology for identifying parameters in a common heat transfer model.

Keywords: nonlinear equations, ordinary differential equations, partial differential equations, Python, non-steady state, heat transfer mechanism

1. Introduction

Numerical methods in heat transfer are essential for solving complex problems that cannot be easily addressed using analytical methods. For example, the Newton–Raphson and the Fixed-Point methods are used to solve equations or systems of algebraic equations that are typically nonlinear. On the other hand, the Finite Difference Method discretizes partial differential equations on a grid of points, facilitating the resolution of heat conduction problems in one, two, or three dimensions. Finally, the Euler and the Runge–Kutta methods are numerical integration techniques used to solve ordinary differential equations, allowing the modeling of the temporal evolution of temperature in dynamic systems. These numerical methods are powerful tools that enable engineers and scientists to obtain approximate and accurate solutions for heat transfer problems in a wide range of applications. This work demonstrates typical heat transfer problems and the algorithms of these numerical methods using Python, a fundamental tool for solving numerical engineering problems. Python offers significant advantages for numerical methods due to its simplicity and ease of use, extensive standard library, and highly active community. Its portability across different platforms and integration with other languages make it versatile. Additionally, Python's productivity-enhancing features and wide-ranging applications make it an ideal choice for implementing numerical methods.

2. Nonlinear equation and system of equations

2.1 Nonlinear equation

One of the problems often encountered in heat transfer is finding roots (x^*) of equations of the form $f(x) = 0$, where $f(x)$ is a real function of a variable x. These functions are either polynomials in x or transcendental functions containing trigonometric, exponential, or logarithmic terms of the independent variable.

For example, in the calculation of steady-state heat conduction in a steam pipe, the problem is to calculate the insulation thickness for a given heat flow. The function, in terms of the diameter (D), that is normally obtained, has the following mathematical structure:

$$f(D) = a\ \ln(D) + \frac{b}{D} + c = 0 \ \ (a, b \text{ and } c \text{ are constant}) \tag{1}$$

There are different numerical methods to find the roots or zeros of these equations. Unfortunately, there is no general algorithm that works for all eqs. [1]. Below, some of the most popular algorithms that allow calculating the real roots of $f(x)$ will be presented.

2.1.1 Fixed-point method or successive substitutions

If we have a general equation $f(x) = 0$, to find a real root, using the fixed-point method, the following steps are applied [1]: 1. This equation is transformed into an algebraic expression of the form $x = g(x)$. 2. An initial value (x_0) is defined. 3. Once x_0 is known, $g(x)$ is evaluated at x_0. The result of this evaluation is denoted as x_1. 4. If $x_0 = x_1$ the process ends, if not, a second evaluation of $g(x)$ is carried out, but now at x_1, the result is denoted as x_2, that is, $g(x_1) = x_2$. 5. This operation is repeated until $x_{n+1} = x_n$ or the difference between them is less than a value set (tolerance, TOL) by the user, that is, $|x_{n+1} - x_n| \leq TOL$.

Example 1: A stainless-steel pipe with a thermal conductivity (k_{steel}) of 50 W/m°C transports saturated steam at 110°C ($T_{\infty 1}$). The outer diameter of the pipe (D_2) is 0.0508 m, while the inner diameter (D_1) is 0.0475 m. The pipe is surrounded by an environment with a temperature of 30°C ($T_{\infty 2}$). The pipe is to be insulated with fiberglass (k_{fiber} =0.045 W/m°C) such that the heat losses (q) are 30 W per linear meter of pipe (L). Calculate the insulation thickness, $((D_3 - D_2)/2)$, required, if the internal ($h_{\infty 1}$) and external ($h_{\infty 2}$) convective heat transfer coefficients are equal to 2000 $W/m^{2o}C$ and 10 $W/m^{2o}C$, respectively.

This problem is solved with the steady-state heat transfer equation (Eq. (2)) considering four (4) thermal resistances: two (2) conductive resistances (pipe and insulating material) and two (2) convective resistances (saturated steam and environment surrounding the pipe).

$$q = \frac{T_{\infty 1} - T_{\infty 2}}{\frac{1}{\pi D_1 L h_{\infty 1}} + \frac{\ln\left(\frac{D_2}{D_1}\right)}{2\pi k_{steel} L} + \frac{\ln\left(\frac{D_3}{D_2}\right)}{2\pi k_{fiber} L} + \frac{1}{\pi D_3 L h_{\infty 2}}} \tag{2}$$

Rearranging (Eq. (2)) gives (Eq. (3)) whose structure is like (Eq. (1)):

$$\left[\frac{1}{2\pi k_{\text{fiber}}}\right]\ln(D_3)+\frac{\left[1/(\pi h_{\infty 2})\right]}{D_3}+\left[\frac{(T_{\infty 2}-T_{\infty 1})}{q'}+\frac{1}{\pi D_1 h_{\infty 1}}+\frac{\ln\left(\frac{D_2}{D_1}\right)}{2\pi k_{\text{steel}}}-\frac{\ln(D_2)}{2\pi k_{\text{fiber}}}\right]=0 \tag{3}$$

where: $q'=\frac{q}{L.}$

To apply the fixed-point method to (Eq. (3)), there are two options to solve for D_3, which are shown in (Eq. (4)) and (Eq. (5)) and obtained by simply solving eq. 3. The method can be applied to both equations. What is not guaranteed is that the method will converge for both mathematical structures.

$$D_3=-\frac{b}{a\ \ln(D_3)+c} \tag{4}$$

$$D_3=e^{-\frac{1}{a}\left(\frac{b}{D_3}+c\right)} \tag{5}$$

where:

$a=\frac{q'}{2\pi k_{\text{fiber}}}$, $b=\frac{q'}{\pi h_{\infty 2}}$, $c=(T_{\infty 2}-T_{\infty 1})+\frac{q'}{\pi D_1 h_{\infty 1}}+\frac{q'\ \ln\left(\frac{D_2}{D_1}\right)}{2\pi k_{\text{steel}}}-\frac{q'\ \ln(D_2)}{2\pi k_{\text{fiber.}}}$

To solve the eqs. (4) and (5), we start with an initial value, which intuitively would be the value of D_2. Possibly, of Eq. (4) and Eq. (5), only one of them converges to a real value, or none of them converges to a real solution, or both converge to real solutions, in any case, these values must be analyzed. The algorithm, programmed in Python and solved in Google Colab (https://colab.research.google.com), is shown below for both equations (**Table 1**).

For (Eq. (4)):	**For (Eq. (5)):**
#Start: Fixed-Point Method	#Start: Fixed-Point Method
import numpy as np # Importing the numpy library for numerical operations	import numpy as np # Importing the numpy library for numerical operations
import matplotlib.pyplot as plt # Importing the matplotlib library for plotting	import matplotlib.pyplot as plt # Importing the matplotlib library for plotting
# Inputs	# Inputs
D1 = 0.0475 # Internal diameter of the pipe [m]	D1 = 0.0475 # Internal diameter of the pipe [m]
D2 = 0.0508 # External diameter of the pipe [m]	D2 = 0.0508 # External diameter of the pipe [m]
Ta1 = 110 # Temperature of saturated steam [°C]	Ta1 = 110 # Temperature of saturated steam [°C]
Ta2 = 30 # Ambient temperature [°C]	Ta2 = 30 # Ambient temperature [°C]
k_steel = 50 # Thermal conductivity of steel [W/m°C]	k_steel = 50 # Thermal conductivity of steel [W/m°C]
k_fiber = 0.045 # Thermal conductivity of fiberglass [W/m°C]	k_fiber = 0.045 # Thermal conductivity of fiberglass [W/m°C]
ha1 = 2000 # Internal convective coefficient [W/m^2° C]	ha1 = 2000 # Internal convective coefficient [W/m^2° C]
ha2 = 10 # External convective coefficient [W/m^2° C]	ha2 = 10 # External convective coefficient [W/m^2° C]
q_L = 30 # Heat flow per unit length of pipe [W/m]	q_L = 30 # Heat flow per unit length of pipe [W/m]
# Initial value of D3 [m]	# Initial value of D3 [m]
D30 = D2	D30 = D2

For (Eq. (4)):	For (Eq. (5)):

# Calculations of a, b, and c 	
a = q_L / (2 * np.pi * k_fiber)
b = q_L / (np.pi * ha2)
c = (Ta2 - Ta1) + q_L / (np.pi * D1 * ha1) + q_L * np.log (D2 / D1) / (2 * np.pi * k_steel) - q_L * np.log(D2) / (2 * np.pi * k_fiber)
 	
# Number of iterations for calculating the root of f (D3) 	
iterations = 2000
D3_values = [] # List to store D3 values for each iteration
 	
for i in range(iterations):
D3 = −b / (a * np.log(D30) + c) # Calculate new D3 value
D3_values.append(D3) # Store the D3 value
D30 = D3 # Update D30 for the next iteration
 	
# Display the value of the real root of f(D3) 	
print(‘The value of D3 [m] is:’)
print(D3)
 	
# Plot the convergence 	
plt.plot(range(iterations), D3_values, marker = ‘o’)
plt.xlabel(‘Iterations’) # Label for the x-axis
plt.ylabel(‘Value of D3 [m]’) # Label for the y-axis
plt.title(‘Convergence of Fixed-Point Method’) # Title of the plot
plt.grid(True) # Enable grid
plt.show()
	# Calculations of a, b, and c
	a = q_L / (2 * np.pi * k_fiber)
	b = q_L / (np.pi * ha2)
	c = (Ta2 - Ta1) + q_L / (np.pi * D1 * ha1) + q_L * np.log (D2 / D1) / (2 * np.pi * k_steel) - q_L * np.log(D2) / (2 * np.pi * k_fiber)
	
	# Number of iterations for calculating the root of f (D3)
	iterations = 2000
	D3_values = [] # List to store D3 values for each iteration
	
	# Iteration for calculating the root of f(D3)
	for i in range(iterations):
	D3 = np.exp.(−(1/a)*(b/D30 + c)) # Calculate new D3 value using fixed-point method
	D3_values.append(D3) # Store the D3 value
	D30 = D3 # Update D30 for the next iteration
	
	# Display the value of the real root of f(D3)
	print(‘The value of D3 [m] is:’)
	print(D3)
	
	# Plot the convergence
	plt.plot(range(iterations), D3_values, marker = ‘o’)
	plt.xlabel(‘Iterations’) # Label for the x-axis
	plt.ylabel(‘Value of D3 [m]’) # Label for the y-axis
	plt.title(‘Convergence of Fixed Point Method’) # Title of the plot
	plt.grid(True) # Enable grid
	plt.show()

Python returns the following solution: The value of D3 [m] is: 0.002352770727297743	Python returns the following solution: The value of D3 [m] is: 0.09843970725382402

Table 1.
Fixed point method codes for Eq. (4) and Eq. (5), respectively.

Figure 1 shows the convergence of the fixed-point method applied to Eq. (4) and **Figure 2** shows the convergence of the fixed-point method applied to Eq. (5); the method converges in a few iterations.

The two previous results converge on real solutions, however, the first solution is not a logical result for the physical description of the pipe, since $D_3 < D_2$. Therefore, the second solution of D_3 would be the indicated answer to obtain the required insulation thickness: Thickness $= {(D_3 - D_2)}/{2} = 0.02382$ m (2.38 cm).

The main limitation of the fixed-point method is its convergence. This method requires that the iteration function is appropriately chosen and meets certain conditions, such as continuity and the existence of a fixed-point. If the function is not contractive in the interval of interest, the method may not converge to a solution, or it may converge very slowly. Additionally, the fixed-point method only guarantees convergence to a single solution, making it less suitable for problems with multiple solutions [1, 2].

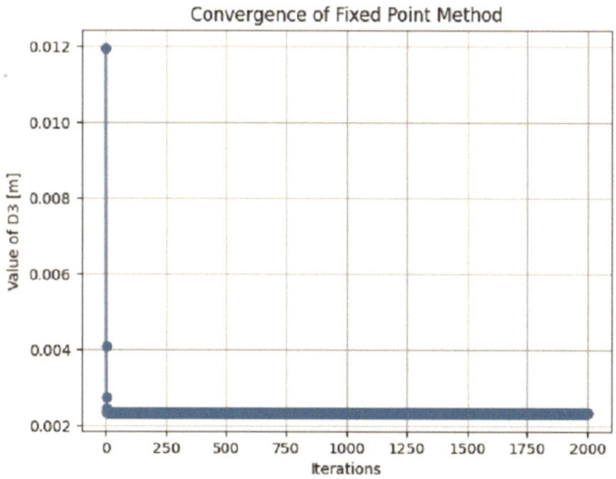

Figure 1.
Convergence of the fixed-point method applied to Eq. (4).

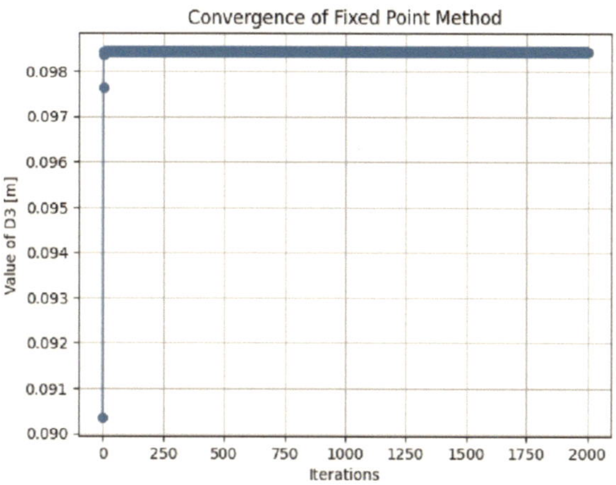

Figure 2.
Convergence of the fixed-point method applied to Eq. (5).

2.1.2 Newton-Raphson method

The Newton-Raphson algorithm is

$$x_{i+1} = x_i - \frac{f(x_i)}{f'(x_i)} \tag{6}$$

With example 1, the Newton-Raphson method is applied, for this case, (Eq. (1)) is:

$$f(D_3) = a\ln(D_3) + \frac{b}{D_3} + c = 0 \tag{7}$$

This (Eq. (7)) has a derivative equal to:

$$f'(D_3) = \frac{a}{D_3} - \frac{b}{{D_3}^2} \tag{8}$$

Therefore, applying the Newton-Raphson algorithm we have:

$$D_{3i+1} = D_{3i} - \frac{f(D_{3i})}{f'(D_{3i})} \tag{9}$$

To calculate the value of the diameter D_3, we start with the initial value D_2 as was done in the fixed-point method.

```
# Start: Newton-Raphson Method
import numpy as np # Importing the numpy library for numerical operations
import matplotlib.pyplot as plt # Importing the matplotlib library for plotting

# Inputs
D1 = 0.0475 # Internal diameter of the pipe [m]
D2 = 0.0508 # External diameter of the pipe [m]
Ta1 = 110 # Temperature of saturated steam [°C]
Ta2 = 30 # Ambient temperature [°C]
k_steel = 50 # Thermal conductivity of steel [W/mK]
k_fiber = 0.045 # Thermal conductivity of fiberglass [W/mK]
ha1 = 2000 # Internal convective coefficient [W/m^2K]
ha2 = 10 # External convective coefficient [W/m^2K]
q_L = 30 # Heat flux per unit length of pipe [W/m]

# Initial value of D3 [m]
D3i = D2

# Calculations of a, b, and c
a = q_L / (2 * np.pi * k_fiber)
b = q_L / (np.pi * ha2)
c = (Ta2 - Ta1) + q_L / (np.pi * D1 * ha1) + q_L * np.log(D2 / D1) / (2 * np.pi * k_steel) - q_L * np.log(D2) /
(2 * np.pi * k_fiber)

# For loop for Newton's Method
iterations = 100
# The number of iterations in a numerical method is chosen by balancing desired accuracy with
computational efficiency.
#A convergence criterion stops iterations when the change between successive solutions is below a defined
tolerance.
#Additionally, a maximum number of iterations is set to prevent infinite loops if convergence is not
achieved, based on experience and preliminary tests.
D3_values = [] # List to store D3 values for each iteration

for i in range(iterations):
  yi = a * np.log(D3i) + b / D3i + c # Calculate the value of the function at D3i
  dyi = a / D3i - b / D3i**2 # Calculate the derivative of the function at D3i
  D3i = D3i - yi / dyi # Newton's Method equation to update D3i
  D3_values.append(D3i) # Store the D3 value
  if abs(yi) <= 0.000001: # Check for convergence.
    #A convergence stopping criterion in a numerical method is usually defined as a condition that
    evaluates whether the differences
#between successive iterations are sufficiently small. A common criterion is to compare the norm of the
difference between the
```

```
#solutions of consecutive iterations with a small threshold (tolerance).
    D3 = D3i # Assign the converged value to D3
    break

# Display the value of the real root of f(D3)
print('The value of D3 [m] is:')
print(D3)

# Plot the convergence
plt.plot(range(len(D3_values)), D3_values, marker='o')
plt.xlabel('Iterations') # Label for the x-axis
plt.ylabel('Value of D3 [m]') # Label for the y-axis
plt.title('Convergence of Newton-Raphson Method') # Title of the plot
plt.grid(True) # Enable grid
plt.show()
# End
```

Python returns the following response: The value of D3 [m] is: 0.098439707253824

Figure 3 shows the convergence of the Newton-Raphson method applied to Eq. (7); the method converges in four iterations, less than in the fixed-point method.

The main limitation of the Newton-Raphson method is its dependence on a good initial guess. If the initial guess is not close to the actual solution, the method may not converge or may converge on an incorrect solution. Additionally, the method can fail if the derivative of the function is zero or very small at any point in the iteration, as this can cause the method to diverge. It is also sensitive to functions with multiple roots and may oscillate or fail to converge if the function is not sufficiently smooth. When the Newton-Raphson method converges, results are obtained in relatively few iterations. However, sometimes the Newton-Raphson method does not converge, this can occur if there is no real root, if the root is an inflection point, or if the initial value is very far from the root wanted and some other region of the function traps the iteration [1, 2].

Nevertheless, the Newton-Raphson method offers several advantages over the fixed-point method, including faster convergence, less dependence on the initial guess, and applicability to a wider range of nonlinear functions. While the fixed-point method requires careful selection of the iteration function and often converges slowly,

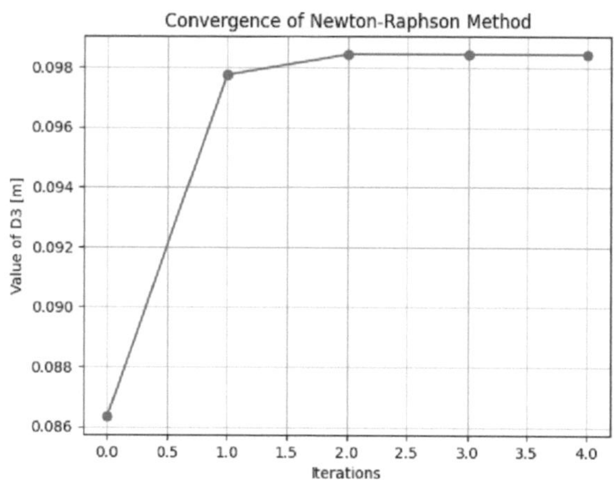

Figure 3.
Convergence of the Newton-Raphson method applied to Eq. (7).

Newton-Raphson's use of the function's derivative allows for more precise approximations and quicker convergence. Additionally, Newton-Raphson can handle complex functions and systems of nonlinear equations, improving stability and convergence rates compared to the basic fixed-point method [2].

2.1.3 Nonlinear equations. Modified Newton-Raphson method through the fsolve *function*

The Python function *fsolve* uses the modified Newton-Raphson method to find the roots of a function. This iterative method employs both the function and its derivative to locate the zeros of a nonlinear function *fsolve* is robust and capable of handling systems of nonlinear equations and unidimensional functions. The modified Newton-Raphson method, an enhanced version of the basic method, incorporates additional techniques to improve convergence and robustness. Common modifications include adjusting the derivative calculation or using advanced numerical methods to address issues when the derivative is small or zero. Another modification involves using a relaxation factor, which reduces the iteration step size to prevent oscillations or divergence. The modified Newton-Raphson method is particularly useful for functions with constraints or complex behavior, as it improves stability and convergence rates compared to the basic method [1, 2].

The *fsolve* function is used to find the root of a transcendental equation $f(x) = 0$. The simplest syntax for this function is *fsolve(fun, x0)*, where *fun* is the name of the function whose roots are being determined and *x0* is an initial value, as close to the root as possible [3]. Below is the Python code for applying the *fsolve* function and Python returns the following response: The root of f(D3) in [m] is: 0.098440.

```
#Start fsolve function
# Import necessary libraries
import numpy as np
from scipy.optimize import fsolve
import matplotlib.pyplot as plt

# Inputs
D1 = 0.0475     # Internal diameter of the pipe [m]
D2 = 0.0508     # External diameter of the pipe [m]
Ta1 = 110      # Steam temperature [°C]
Ta2 = 30     # Ambient temperature [°C]
k_steel = 50   # Thermal conductivity of steel [W/mK]
k_fiber = 0.045  # Thermal conductivity of fiberglass [W/mK]
ha1 = 2000     # Internal convective coefficient [W/m^2K]
ha2 = 10     # External convective coefficient [W/m^2K]
q_L = 30     # Heat flux per unit length of pipe [W/m]

# Calculations of a, b, and c
a = q_L / (2 * np.pi * k_fiber)
b = q_L / (np.pi * ha2)
c = (Ta2 - Ta1) + q_L / (np.pi * D1 * ha1) + q_L * np.log(D2 / D1) / (2 * np.pi * k_steel) - q_L * np.log(D2) /
(2 * np.pi * k_fiber)

# Define the function whose root you want to find
def f(D3):
   return a * np.log(D3) + b / D3 + c

# Initial value of D3 [m]
D30 = D2
```

```
# List to store D3 values for each iteration
D3_values = []

# Define the function to track iterations
def iter_function(D3):
  D3_values.append(D3[0])
  return f(D3)

# Finding the root using fsolve with full_output option to track iterations
root = fsolve(iter_function, D30, full_output=False)

# Display the value of the root
print(f'The root of f(D3) in [m] is: {root[0]:.6f}')

# Plot the convergence
plt.plot(range(len(D3_values)), D3_values, marker='o')
plt.xlabel('Iterations') # Label for the x-axis
plt.ylabel('Value of D3 [m]') # Label for the y-axis
plt.title('Convergence of fsolve Method') # Title of the plot
plt.grid(True) # Enable grid
plt.show()
#End
```

Figure 4 shows the convergence of the *fsolve* function applied to Eq. (7); the method converges in six iterations, less than in the fixed-point method.

2.2 System of equations

A system of several equations with several unknowns can be represented like this:

$$\begin{aligned} f_1(x_1, x_2, x_3, \dots, x_n) &= 0 \\ f_2(x_1, x_2, x_3, \dots, x_n) &= 0 \\ f_n(x_1, x_2, x_3, \dots, x_n) &= 0 \end{aligned} \tag{10}$$

where each of the functions $f_i(x_1, x_2, x_3, \dots, x_n)$, for $1 \le i \le n$, is linear or nonlinear.

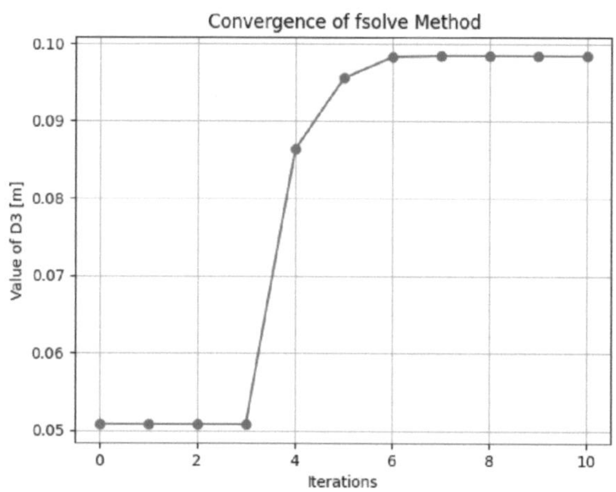

Figure 4.
Convergence of the fsolve *function method applied to Eq. (7).*

In this section, these systems will only be solved with the Python fsolve function. For this purpose, the methodology will be illustrated with the following example.

Example 2: Hot gases from a combustion furnace flow through a square chimney made of concrete ($k_{concrete}$ =1.4 W/m°C). The flow section of the chimney is 0.2 m x 0.2 m, and the wall thickness is 0.2 m. The average temperature of the hot gases in the chimney is Ti = 300°C and the average convection heat transfer coefficient inside the chimney is hi = 70 W/m^2°C. The chimney loses heat from its outer surface to the ambient air at To = 20°C by convection with a heat transfer coefficient of ho = 21 W/m^2 °C and to the sky by radiation. The emissivity of the outer surface of the wall is ϵ_{wall}=0.9 and the effective sky temperature is estimated to be 260 K. Using the finite difference method, with $\Delta x = \Delta y = 0.1$ m and taking full advantage of symmetry, determine the temperatures at nodal points of a cross-section and the rate of heat loss for a 1 m long section of the chimney.

The solution to this problem is given in Ref. [4]. The system of nonlinear equations obtained by applying the finite difference methodology is reproduced here:

$$\begin{aligned}
T_1 - \frac{(T_2 + T_3 + 2865)}{7} &= 0 \\
T_2 - \frac{(T_1 + 2T_4 + 2865)}{8} &= 0 \\
T_3 - \frac{(T_1 + 2T_4 + T_6)}{4} &= 0 \\
T_4 - \frac{(T_2 + T_3 + T_5 + T_7)}{4} &= 0 \\
T_5 - \frac{(T_4 + T_8)}{2} &= 0 \\
T_6 - \frac{(T_2 + T_3 + 456.2 - 0.3645x10^{-9}T_6^4)}{3.5} &= 0 \\
T_7 - \frac{(2T_4 + T_6 + T_8 + 912.4 - 0.729x10^{-9}T_7^4)}{7} &= 0 \\
T_8 - \frac{(2T_5 + T_7 + T_9 + 912.4 - 0.729x10^{-9}T_8^4)}{7} &= 0 \\
T_9 - \frac{(T_8 + 456.2 - 0.364x10^{-9}T_9^4)}{2.5} &= 0
\end{aligned} \tag{11}$$

where T_1, T_2, ..., T_9 are the values of the temperatures at the nodes of the mesh defined for solving the problem.

The Python file for calculating the roots of this system of equations is:

```
#Start: systems of linear or nonlinear equations
# Import necessary libraries

import numpy as np
from scipy.optimize import fsolve
```

```
# Define the system of equations as a function
def F(x):
    return [
        x[0] - (x[1] + x[2] + 2865) / 7, # Equation 1
        x[1] - (x[0] + 2 * x[3] + 2865) / 8, # Equation 2
        x[2] - (x[0] + 2 * x[3] + x[5]) / 4, # Equation 3
        x[3] - (x[1] + x[2] + x[4] + x[6]) / 4, # Equation 4
        x[4] - (x[3] + x[7]) / 2, # Equation 5
        x[5] - (x[1] + x[2] + 456.2 - 0.3645e-9 * x[5]**4) / 3.5, # Equation 6
        x[6] - (2 * x[3] + x[5] + x[7] + 912.4 - 0.729e-9 * x[6]**4) / 7, # Equation 7
        x[7] - (2 * x[4] + x[6] + x[8] + 912.4 - 0.729e-9 * x[7]**4) / 7, # Equation 8
        x[8] - (x[7] + 456.2 - 0.364e-9 * x[8]**4) / 2.5, # Equation 9
    ]

# Initial guess for the variables (temperature values T(i))
x0 = [300, 300, 300, 300, 300, 300, 300, 300, 300]

# Solve the system of equations using fsolve
x, infodict, ier, mesg = fsolve(F, x0, full_output=True)

# Define the variable names for the solution
solution = ['T1', 'T2', 'T3', 'T4', 'T5', 'T6', 'T7', 'T8', 'T9']

# Store the solution values
values = x

# Calculate the residues
residues = F(x)

# Show results using structured labels
print('Answer:')
for var, val, res in zip(solution, values, residues):
    print(f'{var} = {val:.6f} °C, Residue = {res:.6e}') # Print each variable with its value
    and residue

# Print temperature values in a structured format
print(f'\nThe temperature values are:')
for var, val in zip(solution, values):
    print(f'{var}: {val:.2f} °C')

# Show convergence information
if ier == 1:
    print(f"\nConverged successfully in {infodict['nfev']} function evaluations.") # Print
    success message
else:
    print("\nThe solution did not converge.")
#End
```

Answer:
T1 = 551.208098°C, Residue = 1.136868e-13
T2 = 535.634737°C, Residue = 0.000000e+00
T3 = 457.821950°C, Residue = −5.684342e-14
T4 = 434.434898°C, Residue = 5.684342e-14
T5 = 384.899584°C, Residue = 0.000000e+00
T6 = 411.209905°C, Residue = −7.668177e-11
T7 = 359.383322°C, Residue = −5.798029e-12
T8 = 335.364269°C, Residue = 3.677769e-11
T9 = 315.188749°C, Residue = 1.142553e-11

The temperature values are:
T1: 551.21°C
T2: 535.63°C
T3: 457.82°C
T4: 434.43°C
T5: 384.90°C
T6: 411.21°C
T7: 359.38°C
T8: 335.36°C
T9: 315.19°C

Converged successfully in 15 function evaluations.

2.3 Ordinary differential equations (ODE). Numerical solution of ODE, initial value problem

A differential equation contains a dependent variable and its derivatives with respect to one or more independent variables. Many differential equations arise from applying the principles of conservation of mass and energy in engineering processes [5]. In this section, ordinary differential equations are solved with the initial value problem by numerical methods. To illustrate this, the following problem is presented:

Example 3: A metal sphere is surrounded by a convective environment and its initial temperature is very high (lower than the melting point of the metal) such that radiation is considered. The sphere has a homogeneous initial temperature $T_{(0)} = T_i$ and will be cooled in contact with air that is at the temperature T_∞. The mathematical model of heat transfer in a non-steady state, if no appreciable temperature gradients occur within the sphere, is as follows [4]:

$$\rho V c_p \frac{dT_{(t)}}{dt} = -hA_s\left(T_{(t)} - T_\infty\right) - \epsilon\sigma A_s\left({T_{(t)}}^4 - {T_{surroundings}}^4\right) \tag{12}$$

Assuming the surroundings are in equilibrium with the air surrounding the sphere, then $T_\infty = T_{surroundings}$, therefore:

$$\rho V c_p \frac{dT_{(t)}}{dt} = -hA_s\left(T_{(t)} - T_\infty\right) - \epsilon\sigma A_s\left({T_{(t)}}^4 - {T_\infty}^4\right) \tag{13}$$

Eq. (13) is a nonlinear first-order ordinary differential equation, which makes finding an analytical solution, either extremely difficult or impossible with existing transcendental functions. Therefore, we rely on numerical methods to solve it.

This problem will be solved with the following values: $\rho = 8933\ kg/m^3$; $V = 4.189x10^{-6}\ m^3$; $c_p = 385\ J/(kg\ K)$; $h = 10\ W/m^2K$; $A_s = 0.00126\ m^2$; $T_\infty = 30^\circ C = 303{,}15\ K$; $\epsilon_{sphere} = 0.15$; $\sigma = 5.67\ x\ 10^{-8}\ \frac{W}{m^2K^4}$ (Stefan – Boltzmann constant). The initial value of Temperature is: $T_{(0)} = T_i = 1000\ K$.

Eq. (14) is derived by substituting all the parameters in Eq. (13) and separating all terms except for the first-order derivative of temperature:

$$\frac{dT_{(t)}}{dt} = 0.271413 - 8.745852x10^{-4}T_{(t)} - 7.4383x10^{-13}{T_{(t)}}^4 \tag{14}$$

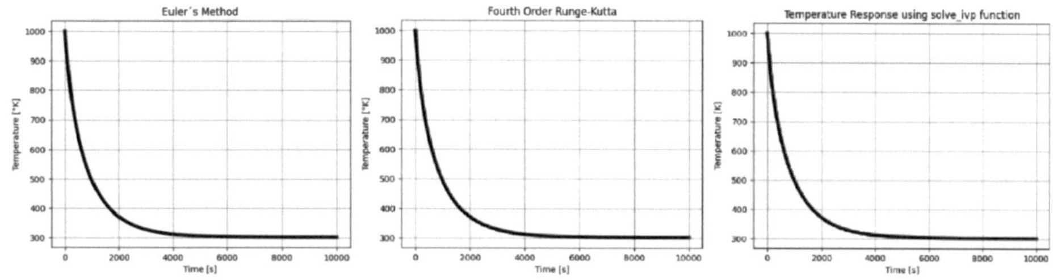

Figure 5.
Temperature response using methods for solving Eq. (17): Euler, Runge-Kutta, and solve_ivp *function.*

The most used numerical methods, Euler and Runge-Kutta, for this problem are outlined below.

2.3.1 Euler's method

This method is the simplest to solve an initial value ODE problem. Euler's algorithm is summarized in (Eq. (15)):

$$y_{n+1} = y_n + h\, f(x_n, y_n) \tag{15}$$

where h is the integration step of the differential equation, $f(x_n, y_n)$ is the evaluation of the derivative at the current instant n, y_n is the value of the dependent variable at the current instant n, and y_{n+1} is the value of the dependent variable at the next instant $n + 1$ [6]. Euler's algorithm for the sphere in example 3 is expressed as follows:

$$T_{n+1} = T_n + h\,\left(0.271413 - 8.745852\text{x}10^{-4} T_n - 7.4383\text{x}10^{-13} {T_n}^4\right) \tag{16}$$

with the initial condition: $T_{(0)} = 1000$ K

The problem will be solved with one step of h=0.01 until the temperature value of the sphere is close to the air temperature (steady state condition). The computational model in Python is as follows and **Figure 5** shows the graphical solution.

```
# Start: Euler's Method
import numpy as np
import matplotlib.pyplot as plt

# Initial condition, step size, and simulation time
t = [0]        # Initialize time array with start time t(0)
T = [1000]        # Initialize temperature array with initial temperature T(0)
h = 0.01        # Step size for the Euler method
tsimula = 10000        # Total simulation time in seconds

# Define the function f, representing dT/dt
def f(t, T):
return 0.271413 - 8.745852e-4 * T - 7.4383e-13 * T**4 # Differential equation

# Simulation loop
for i in range(int(tsimula / h)):
    T.append(T[-1] + h * f(t[-1], T[-1])) # Calculate the next temperature using Euler's algorithm
    t.append(t[-1] + h)        # Update the time vector with the next time step
```

```
# Plot temperature versus time
plt.plot(t, T, '-k', linewidth=4)    # Plot temperature (T) against time (t)
plt.title('Euler\'s Method')         # Set plot title
plt.xlabel('Time [s]')               # Set x-axis label
plt.ylabel('Temperature [°K]')       # Set y-axis label
plt.grid(True)                       # Display grid
plt.show()                           # Show the plot
# End
```

Euler's method, though simple and easy to implement, has significant limitations in terms of accuracy and stability. It is particularly unsuitable for stiff differential equations due to its sensitivity to step size, leading to cumulative errors and inaccurate results. These limitations make more advanced methods like Runge-Kutta preferable for more complex problems [1].

2.3.2 Runge-Kutta method

The Euler method is a straightforward numerical approach for solving ordinary differential equations by taking linear steps based on the slope at the current point, but it can accumulate significant errors over larger intervals due to its simplicity. On the other hand, the fourth-order Runge-Kutta method enhances accuracy by using multiple intermediate points and weighted averages within each step, leading to a much more precise approximation of the solution with less error accumulation. Although the Runge-Kutta method is more computationally demanding, its ability to balance accuracy and efficiency makes it a preferred choice in many scientific and engineering applications over the simpler Euler method [1].

The fourth-order Runge-Kutta algorithm is summarized in (Eq. (17)):

$$y_{n+1} = y_n + \frac{h\,(f_1 + 2f_2 + 2f_3 + f_4)}{6} \tag{17}$$

where:

$$f_1 = f(t_n, y_n)$$

$$f_2 = f\left(t_n + \frac{h}{2}, y_n + \frac{h}{2} f_1\right)$$

$$f_3 = f\left(t_n + \frac{h}{2}, y_n + \frac{h}{2} f_2\right)$$

$$f_4 = f(t_n + h, y_n + hf_3).$$

Furthermore, h is the integration step of the differential equation, $f(x_n, y_n)$ is the evaluation of the derivative at the current time n, y_n is the value of the dependent variable at the current time n and y_{n+1} is the value of the dependent variable at a later time $n + 1$ [1].

Returning to the sphere problem solved in the section on Euler's method, we have the following Python file that applies the fourth-order Runge-Kutta method. Python returns the solution shown in **Figure 5**.

```
# Start: Runge-Kutta Method
import numpy as np
import matplotlib.pyplot as plt

# Initial condition
t = [0] # Initial time [s]
T = [1000] # Initial temperature [K]
h = 0.01 # Step size [s]
tsimula = 10000 # Total simulation time [s]

# Define the function f, representing dT/dt
def f(t, T):
    return 0.271413 - 8.745852e-4 * T - 7.4383e-13 * T**4

# Simulation loop using Fourth Order Runge-Kutta method
for i in range(int(tsimula / h)):
    k1 = f(t[i], T[i]) # Compute first slope (k1)
    k2 = f(t[i] + h / 2, T[i] + h * k1 / 2) # Compute second slope (k2)
    k3 = f(t[i] + h / 2, T[i] + h * k2 / 2) # Compute third slope (k3)
    k4 = f(t[i] + h, T[i] + h * k3) # Compute fourth slope (k4)

    t.append((i + 1) * h) # Increment time by step size
    T.append(T[i] + h * (k1 + 2 * k2 + 2 * k3 + k4) / 6) # Compute next temperature value

# Plot Temperature versus Time
plt.plot(t, T, '-k', linewidth=4)
plt.title('Fourth Order Runge-Kutta')
plt.xlabel('Time [s]')
plt.ylabel('Temperature [K]')
plt.grid(True)
plt.show()
# End
```

For advanced heat transfer applications involving ordinary differential equations (ODEs), there are several alternative numerical solvers to consider, such as the Adams-Bashforth-Moulton Method, which is a predictor-corrector method combining the Adams-Bashforth and Adams-Moulton techniques for improved accuracy; the Backward Differentiation Formulas (BDF), which are implicit methods particularly useful for stiff ODEs; the Bulirsch-Stoer Method, known for its high accuracy through polynomial extrapolation; Gear's Method, specialized for stiff ODEs by adjusting step size and order for stability and accuracy; and the Runge-Kutta-Fehlberg Method, an adaptive Runge-Kutta method with dynamic error control and step size adjustment for efficient and accurate ODE solutions. The reader is invited to consult the different bibliographic sources that discuss these topics [1, 2].

2.3.3 Solve_ivp function

Python has built-in functions such as the *solve_ivp* function (among others). The default numerical method used in Python's *solve_ivp* function is RK45. This method is a variant of the Runge-Kutta method of order 5(4), which is an explicit method for integrating ordinary differential eqs. [3]. *solve_ivp* function is highly popular because of its versatility, providing a unified interface to multiple integration methods for solving ODEs, and its adaptive step size control enhances efficiency and accuracy. In contrast, the fixed-step Runge-Kutta method, while simpler, maintains a constant step size throughout the integration process, which can lead to inefficiencies and larger errors, especially in stiff or complex ODE systems. Despite being computationally lighter, the fixed-step approach lacks the flexibility and adaptive precision of *solve_ivp*, making it less suitable for advanced applications requiring high accuracy [3]. Therefore, another way to solve the above problem is by using this function as shown below:

```
# Start: Numerical Solution using solve_ivp function
import numpy as np
import matplotlib.pyplot as plt
from scipy.integrate import solve_ivp

# Initial condition and simulation time
tsimula = [0, 10000] # Simulation interval [a, b] in seconds
T0 = 1000 # Initial temperature value in Kelvin
# Define the function f, representing dT/dt
def f(t, T):
    return 0.271413 - 8.745852e-4 * T - 7.4383e-13 * T**4
# Solving the system of differential equations using solve_ivp
sol = solve_ivp(f, tsimula, [T0], t_eval=np.linspace(tsimula[0], tsimula[1], 1000))
# Plot the temperature response over time
plt.plot(sol.t, sol.y[0], '-k', linewidth=4)
plt.grid(True)
plt.title('Temperature Response using solve_ivp function')
plt.xlabel('Time [s]') # X-axis label
plt.ylabel('Temperature [K]') # Y-axis label
plt.show()
# End
```

Python returns the solution shown in **Figure 5**.

2.4 System of ordinary differential equations

The methods seen in the previous sections are extended to be applied to systems of ordinary differential equations. Next, through an example, the application of the fourth-order Euler and Runge-Kutta algorithms will be shown, as well as the *solve_ivp* function [3].

Example 4: We have a perfectly mixed jacketed tank where a mass m_t of fluid A is heated. A fluid B enters and exits the jacket with a constant flow. It is assumed that there is no heat loss to the surroundings, the specific heat of the fluids is constant and the temperature of the fluid entering the jacket is constant.

The dynamic model showing the evolution of temperature in fluids A and B is:

$$\frac{dT_{t(t)}}{dt} = \frac{UA\left(T_{c(t)} - T_{t(t)}\right)}{m_t c_{p,t}} \quad \text{(For the fluidA)} \tag{18}$$

$$\frac{dT_{c(t)}}{dt} = \frac{UA\left(T_{t(t)} - T_{c(t)}\right)}{m_c c_{p,c}} + \frac{\dot{m}_c\left(T_{in} - T_{c(t)}\right)}{m_c} \quad \text{(For the fluid B)} \tag{19}$$

where: $T_{t(t)}$: the temperature of the fluid in the tank, [K]; $T_{c(t)}$: the temperature of the fluid in the jacket, [K]; T_{in} : The temperature of the fluid at the jacket inlet, [K];

U :overall heat transfer coefficient [W/m^2K]; A :heat transfer area [m^2]; m_t :mass of fluid in the tank [kg]; m_c :mass of fluid in the jacket [kg]; $\dot{m}_c$:mass flow of fluid in the jacket [kg/s]; $c_{p,t}$: specific heat of fluid in the tank [J/kgK]; $c_{p,c}$:specific heat of fluid in the jacket [J/kgK].

2.4.1 System of ordinary differential equations. Euler's method

Euler's algorithm, for this case, is as follows [6]:

$$y_{n+1} = y_n + h\, f(x_n, y_n, z_n) \tag{20}$$

$$z_{n+1} = z_n + h\, g(x_n, y_n, z_n) \tag{21}$$

The Python file for the proposed problem is:

```
# Start: Euler Method - ODE Systems
import numpy as np
import matplotlib.pyplot as plt

# System parameters
U = 500 # Overall heat transfer coefficient [W/m²·K]
A = 1.0 # Heat transfer area of the jacket [m²]
m_t = 100 # Mass of the fluid in the tank [kg]
C_p_t = 4184 # Specific heat of the fluid in the tank [J/kg·K]
m_c = 50 # Mass of the fluid in the jacket [kg]
C_p_c = 4184 # Specific heat of the fluid in the jacket [J/kg·K]
dot_m = 0.1 # Mass flow rate of the fluid in the jacket [kg/s]
T_in = 373 # Inlet temperature of the fluid to the jacket [K]

# Initial conditions
T_t0 = 300 # Initial temperature of the fluid in the tank [K]
T_c0 = 350 # Initial temperature of the fluid in the jacket [K]

# Simulation time
tsimula = [0, 10000] # Simulation interval [s]
# Define the function f, i.e., dT/dt
def f(t, T):
  T_t, T_c = T
  dT_t_dt = (U * A * (T_c - T_t)) / (m_t * C_p_t) # Rate of change of temperature in the tank
  dT_c_dt = (U * A * (T_t - T_c)) / (m_c * C_p_c) + (dot_m * (T_in - T_c)) / m_c # Rate of change of
  temperature in the jacket. A '*' was added here.
  return [dT_t_dt, dT_c_dt]

# Simulation loop using Euler's method
h = 0.01 # Step size
t = np.arange(tsimula[0], tsimula[1], h) # Time array for the simulation interval
T_t = np.zeros(len(t)) # Array to store temperature in the tank over time
T_c = np.zeros(len(t)) # Array to store temperature in the jacket over time
T_t[0] = T_t0 # Initial temperature of the tank at t=0
T_c[0] = T_c0 # Initial temperature of the jacket at t=0

for i in range(1, len(t)):
  dT_t_dt, dT_c_dt = f(t[i-1], [T_t[i-1], T_c[i-1]]) # Compute the rate of change of temperature at the
  previous step
  T_t[i] = T_t[i-1] + h * dT_t_dt # Update temperature of the tank
  T_c[i] = T_c[i-1] + h * dT_c_dt # Update temperature of the jacket
```

```
# Plot the response
plt.plot(t, T_t, label='T_t (Tank)')
plt.plot(t, T_c, label='T_c (Jacket)')
plt.grid(True)
plt.title('Dynamic Model of a Batch Stirred Tank with External Jacket (Euler´s method)')
plt.xlabel('Time, [s]')
plt.ylabel('Temperature, [K]')
plt.legend()
plt.show()
# End
```

Python returns the solution shown in **Figure 6.**

2.4.2 System of ordinary differential equations. Runge-Kutta method

The fourth-order Runge-Kutta algorithm, for this case, is as follows [6]:

$$y_{n+1} = y_n + \frac{h\,(k_1 + 2k_2 + 2k_3 + k_4)}{6} \tag{22}$$

$$z_{n+1} = z_n + \frac{h\,(l_1 + 2l_2 + 2l_3 + l_4)}{6} \tag{23}$$

where:

$$k_1 = f(t_n, y_n, z_n); \qquad l_1 = g(t_n, y_n, z_n)$$

$$k_2 = f\left(t_n + \frac{h}{2}, y_n + \frac{h}{2}k_1, z_n + \frac{h}{2}l_1\right); \qquad l_2 = g\left(t_n + \frac{h}{2}, y_n + \frac{h}{2}k_1, z_n + \frac{h}{2}l_1\right)$$

$$k_3 = f\left(t_n + \frac{h}{2}, y_n + \frac{h}{2}k_2, z_n + \frac{h}{2}l_2\right); \qquad l_3 = g\left(t_n + \frac{h}{2}, y_n + \frac{h}{2}k_2, z_n + \frac{h}{2}l_2\right)$$

$$k_4 = f(t_n + h, y_n + hk_3, z_n + hl_3); \qquad l_4 = g(t_n + h, y_n + hk_3, z_n + hl_3)$$

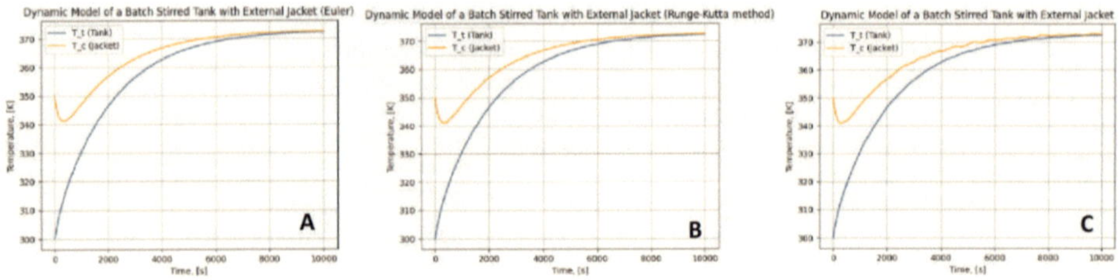

Figure 6.
Temperature response using methods for solving Eqs. (18) and (19): Euler (A), Runge-Kutta (B), and solve_ivp *function (C).*

The Python file to solve the problem is:

```
# Start: Runge-Kutta Method - ODE System
import numpy as np
import matplotlib.pyplot as plt
from scipy.integrate import solve_ivp
# System parameters
U = 500   # Overall heat transfer coefficient [W/m²·K]
A = 1.0   # Heat transfer area of the jacket [m²]
m_t = 100  # Mass of the fluid in the tank [kg]
C_p_t = 4184 # Specific heat of the fluid in the tank [J/kg·K]
m_c = 50   # Mass of the fluid in the jacket [kg]
C_p_c = 4184 # Specific heat of the fluid in the jacket [J/kg·K]
dot_m = 0.1 # Mass flow rate of the fluid in the jacket [kg/s]
T_in = 373 # Inlet temperature of the fluid to the jacket [K]
# Initial conditions and simulation time
T_t0 = 300  # Initial temperature of the fluid in the tank [K]
T_c0 = 350  # Initial temperature of the fluid in the jacket [K]
tsimula = [0, 10000] # Simulation interval [s]
# Define the function f, representing the system of ODEs
def f(t, T):
   T_t, T_c = T # Unpack temperature values for the tank and jacket
   dT_t_dt = (U * A * (T_c - T_t)) / (m_t * C_p_t) # Rate of change of temperature in the tank
   dT_c_dt = (U * A * (T_t - T_c)) / (m_c * C_p_c) + (dot_m * (T_in - T_c)) / m_c # Rate of change of
   temperature in the jacket
   return [dT_t_dt, dT_c_dt] # Return derivatives as a list

# Simulation loop using the fourth-order Runge-Kutta method
h = 0.01 # Step size
t = np.arange(tsimula[0], tsimula[1], h) # Time vector
T_t = np.zeros(len(t)) # Initialize temperature array for the tank
T_c = np.zeros(len(t)) # Initialize temperature array for the jacket
T_t[0] = T_t0 # Set initial temperature of the tank
T_c[0] = T_c0 # Set initial temperature of the jacket

for i in range(1, len(t)):
   k1 = np.array(f(t[i-1], [T_t[i-1], T_c[i-1]])) # Compute first slope (k1)
   k2 = np.array(f(t[i-1] + h/2, [T_t[i-1] + h*k1[0]/2, T_c[i-1] + h*k1[1]/2])) # Compute second slope (k2)
   k3 = np.array(f(t[i-1] + h/2, [T_t[i-1] + h*k2[0]/2, T_c[i-1] + h*k2[1]/2])) # Compute third slope (k3)
   k4 = np.array(f(t[i-1] + h, [T_t[i-1] + h*k3[0], T_c[i-1] + h*k3[1]])) # Compute fourth slope (k4)
   T_t[i] = T_t[i-1] + (h/6) * (k1[0] + 2*k2[0] + 2*k3[0] + k4[0]) # Update temperature
   in the tank
   T_c[i] = T_c[i-1] + (h/6) * (k1[1] + 2*k2[1] + 2*k3[1] + k4[1]) # Update temperature in the jacket

# Plot the response
plt.plot(t, T_t, label='T_t (Tank)')
plt.plot(t, T_c, label='T_c (Jacket)')
plt.grid(True)
plt.title('Dynamic Model of a Batch Stirred Tank with External Jacket (Runge-Kutta method)')
plt.xlabel('Time, [s]')
plt.ylabel('Temperature, [K]')
plt.legend()
plt.show()
# End
```

Python returns the solution shown in **Figure 6**.

2.4.3 System of ordinary differential equations. Solve_ivp *method*

Another way to deal with the above problem is to use the *solve_ivp* function as shown in the following file [3]:

```
# Start: solve_ivp method - ODE system
import numpy as np
import matplotlib.pyplot as plt
from scipy.integrate import solve_ivp
# System parameters
U = 500     # Overall heat transfer coefficient [W/m²·K]
A = 1.0     # Heat transfer area of the jacket [m²]
m_t = 100     # Mass of the fluid in the tank [kg]
C_p_t = 4184   # Specific heat of the fluid in the tank [J/kg·K]
m_c = 50     # Mass of the fluid in the jacket [kg]
C_p_c = 4184   # Specific heat of the fluid in the jacket [J/kg·K]
dot_m = 0.1   # Mass flow rate of the fluid in the jacket [kg/s]
T_in = 373    # Inlet temperature of the fluid to the jacket [K]
# Initial conditions
T_t0 = 300    # Initial temperature of the fluid in the tank [K]
T_c0 = 350    # Initial temperature of the fluid in the jacket [K]
# Simulation time
tsimula = [0, 10000] # Simulation interval [s]
# Define the function f, i.e., dT/dt
def f(t, T):
T_t, T_c = T
dT_t_dt = (U * A * (T_c - T_t)) / (m_t * C_p_t)
dT_c_dt = (U * A * (T_t - T_c)) / (m_c * C_p_c) + (dot_m * (T_in - T_c)) / m_c
return [dT_t_dt, dT_c_dt]
# Solve the system of differential equations
sol = solve_ivp(f, tsimula, [T_t0, T_c0], t_eval=np.linspace(tsimula[0], tsimula[1], 1000))
# t_eval specifies the time points where the solution is computed.
# It's an array of times at which the solution is desired,
# which allows for more precise control over the output intervals.

# Plot the response
plt.plot(sol.t, sol.y[0], label='T_t (Tank)')
plt.plot(sol.t, sol.y[1], label='T_c (Jacket)')
plt.grid(True)
plt.title('Dynamic Model of a Batch Stirred Tank with External Jacket')
plt.xlabel('Time, [s]')
plt.ylabel('Temperature, [K]')
plt.legend()
plt.show()
# End
```

Python returns the solution shown in **Figure 6**.

2.5 Partial differential equations (PDEs)

Differential equations are formed by partial derivatives of functions that depend on several variables. Despite the complexity of PDEs, there are numerical methods proposed to solve cases that are very useful in engineering [5]. In this section, the cases of parabolic PDEs and elliptic PDEs will be discussed.

If the PDE is the second order of the form

$$a_1 \frac{\partial^2 u}{\partial y^2} + 2a_2 \frac{\partial^2 u}{\partial x \partial y} + a_3 \frac{\partial^2 u}{\partial x^2} + a_4 \frac{\partial u}{\partial x} + a_5 \frac{\partial u}{\partial y} + a_6 u + a_7 = 0 \tag{24}$$

is classified as a parabolic PDE if the value of the discriminant ${a_2}^2 - a_1 a_3 = 0$ and elliptic PDE if the value of the discriminant ${a_2}^2 - a_1 a_3 < 0$.

2.5.1 PDE: Solutions by the finite difference method (FDM)

The Finite Difference Method (FDM) is widely used due to its simplicity, grid-based approach, flexibility, and computational efficiency. It is easy to understand and implement, making it accessible for various users. The grid-based approach is intuitive for solving partial differential equations, and FDM can adapt to different problems and boundary conditions. Additionally, FDM provides a balance between accuracy and computational cost, and its extensive use in practice means there is a wealth of resources and literature available.

The Finite Difference Method (FDM) should be used over the Finite Element Method (FEM) and Finite Volume Method (FVM) when simplicity and computational efficiency are key considerations, and the problem involves regular geometries with structured grids. FDM is particularly suitable for solving partial differential equations (PDEs) in problems such as heat conduction, fluid flow, and wave propagation, where the domain and boundary conditions are straightforward. Unlike FEM and FVM, FDM is easier to implement, especially for problems with simple boundary conditions, making it a preferred choice for quick, less complex analyses. However, FDM is limited to structured grids and may not handle complex geometries as effectively as FEM and FVM, which offer greater flexibility for irregular meshes and conservation law applications.

The finite difference method consists of replacing the numerical derivation formulas in the differential equation to be solved. The method consists of constructing a recurrence equation that through a pivoting process will provide a series of equations that will be solved by a specific method [5].

2.5.2 Parabolic PDE

To apply the FDM in a parabolic PDE, the one-dimensional heat transfer equation shown below will be taken [4]:

$$\frac{\partial^2 u}{\partial x^2} = \frac{1}{\alpha} \frac{\partial u}{\partial t} \tag{25}$$

on the domain $0 \le t \le t_f$ and $0 \le x \le 1$.

The initial and boundary conditions are: $u(x,0) = f(x), (0 \le x \le 1);\ u(0,t) = g_0(t)$, $u(1,t) = g_1(t), (0 \le t \le t_f)$.

To create a grid in the FDM, first define the problem domain and discretize it into a grid by dividing the domain into equal intervals. In one spatial dimension, this involves dividing the interval $[x_0, x_f]$ into n subintervals, creating n + 1 grid points where solutions will be calculated. The distance between each grid point, or increment, is denoted as Δx. If the problem also depends on time, select a time step Δt and divide the time interval $[t_0, t_f]$ into m subintervals, creating m + 1 time points, as shown in **Figure 7**. This discretization converts differential equations into algebraic equations that can be solved at each grid point, allowing the problem's solution to be approximated across the entire domain.

To apply the FDM to the Eq. (25), the domain t is divided into m sections and the domain x into n sections such that $\Delta t = {}^{t_f}/_{m}$ and $\Delta x = {}^{1}/_{n}$. Therefore, the interior points are given by $t_j = j\Delta t$ $(j = 1, 2, \ldots, m-1)$ and $x_i = i\Delta x$ $(i = 1, 2, \ldots, n-1)$. The solution at point (x_i, t_j) is represented as $u_{i,j}$. The points involved in the calculations at time steps $j-1$, j and $j+1$ are shown in **Figure 7**.

Partial differential terms can be approximated by finite differences such as:

$$\frac{\partial u}{\partial t} \cong \frac{1}{\Delta t}\left(u_{i,j+1} - u_{i,j}\right) \tag{26}$$

$$\frac{\partial^2 u}{\partial x^2} \cong \frac{1}{(\Delta x)^2}\left(u_{i-1,j} - 2u_{i,j} + u_{i+1,j}\right) \tag{27}$$

By replacing the partial differential terms by the finite differences, we obtain a linear system of equations for u at the node points:

$$u_{i,j+1} - u_{i,j} = \frac{\alpha \Delta t}{(\Delta x)^2}\left(u_{i-1,j} - 2u_{i,j} + u_{i+1,j}\right) = r\left(u_{i-1,j} - 2u_{i,j} + u_{i+1,j}\right) \tag{28}$$

where $r = \frac{\alpha \Delta t}{(\Delta x)^2}$. Solving for $u_{i,j+1}$, we have

$$u_{i,j+1} = r\,u_{i-1,j} + (1-2r)u_{i,j} + ru_{i+1,j} \quad (i = 2, 3, \ldots, n-2) \tag{29}$$

$$u_{1,j+1} = r\,g_{0,j} + (1-2r)u_{1,j} + ru_{2,j} \quad (i = 1) \tag{30}$$

$$u_{n-1,j+1} = r\,u_{n-2,j} + (1-2r)u_{n-1,j} + rg_{1,j} \quad (i = n-1) \tag{31}$$

The above procedure is known as an explicit scheme. For the solution to be stable, it is required that $r \leq 0.5$.

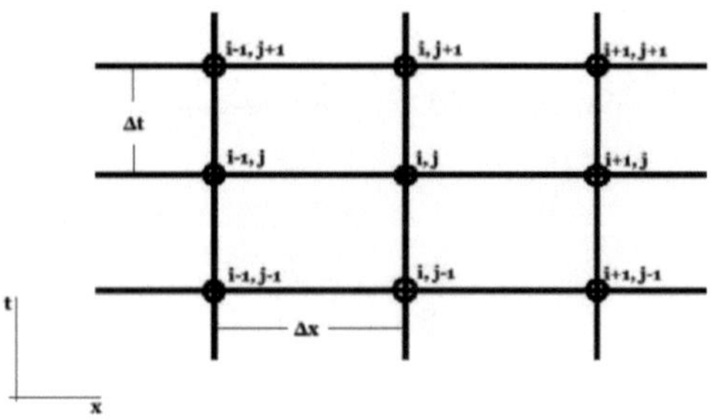

Figure 7.
Mesh of the domain of x and t.

Example 5: The temperature distribution in a bar of unit length is given by:

$$\frac{\partial^2 T}{\partial x^2} = \frac{1}{\alpha}\frac{\partial T}{\partial t} \quad \begin{matrix} 0 \leq x \leq 1) \\ (0 \leq t \leq t_f, \end{matrix} \tag{32}$$

The initial condition is given by $T(x, 0) = 120^{\circ}C$ and the ends are kept at 20°C on the left side and $0^{\circ}C$ on the right side of the bar, that is, $T(0, t) = 20$ and y $T(1, t) = 0$ $(0 \leq t \leq t_f)$. Make the temperature profile graph on the bar for: $t_f = 750$, $\alpha = 1.0x10^{-4}\ m^2/s$, $m = 80$, $n = 1000$. The Python file to solve the one-dimensional heat conduction equation with these values is given below:

```
# Start: Parabolic PDE
import numpy as np
import matplotlib.pyplot as plt
from mpl_toolkits.mplot3d import Axes3D

def parabolic_pde(f, g0, g1, tf, n_x, n_t, alpha):
  # Inputs
  h = 1 / n_x # Length of subintervals in x
  d = tf / n_t # Length of subintervals in t
  r = alpha * d / h**2 # Stability parameter r, should be less than 0.5
  print(f'The value of r is: {r}') # Print the value of r

  # Vectors of variables x and t
  x = np.linspace(0, 1, n_x + 1)
  t = np.linspace(0, tf, n_t + 1)

  # Initial and boundary conditions
  T = np.zeros((n_t + 1, n_x + 1))
  T[0, :] = f(x) # Initial condition: T(x,0)=f(x)
  T[:, 0] = g0(t) # Boundary condition at x=0: T(0,t)=g0(t)
  T[:, -1] = g1(t) # Boundary condition at x=1: T(1,t)=g1(t)

# Check stability condition
if r >= 0.5: # Ensure that the stability condition is satisfied
  raise ValueError('r must be <= 0.5')

# Solve the PDE using the finite differences explicit method
for k in range(n_t):
  T[k + 1, 1:-1] = r * T[k, :-2] + (1 - 2 * r) * T[k, 1:-1] + r * T[k, 2:]

  # Plot results in 3D
  X, Y = np.meshgrid(x, t)
  fig = plt.figure()
  ax = fig.add_subplot(111, projection='3d')
  surf = ax.plot_surface(X, Y, T, cmap='viridis')

  # Add color bar to show temperature scale
  cbar = fig.colorbar(surf, ax=ax, shrink=0.5, aspect=5)
  cbar.ax.set_ylabel('Temperature [°C]', fontsize=10)

  ax.set_xlabel('Length [m], x', fontsize=10)
  ax.set_ylabel('Time [s], t', fontsize=10)
  ax.set_zlabel('Temperature [°C], T(t,x)', fontsize=10)
```

```
    plt.tight_layout() # Adjust margins for better layout
    plt.show()

    # Plot contour results
    fig2, ax2 = plt.subplots()
    CS = ax2.contour(X, Y, T, levels=6)
    ax2.clabel(CS, inline=True, fontsize=10)

    # Add color bar to contour plot
    cbar2 = plt.colorbar(CS, ax=ax2, shrink=0.8)
    cbar2.ax.set_ylabel('Temperature [°C]', fontsize=10)

    ax2.set_xlabel('Length [m], x', fontsize=10)
    ax2.set_ylabel('Time [s], t', fontsize=10)
    plt.tight_layout() # Adjust margins for better layout
    plt.show()

    return T

# Define functions for initial and boundary conditions
f = lambda x: 120 # Initial condition: T(x,0)=f(x)
g0 = lambda t: 20 # Boundary condition at x=0: T(0,t)=g0(t)
g1 = lambda t: 0 # Boundary condition at x=1: T(1,t)=g1(t)

# Problem parameters
n_x = 80 # Number of subintervals in x
n_t = 1000 # Number of subintervals in t
tf = 750 # Final time
alpha = 1.0e-4 # Thermal diffusivity [m^2/s]

# Call the function to solve the parabolic PDE
T = parabolic_pde(f, g0, g1, tf, n_x, n_t, alpha)
#End
```

Python returns the value of r: The value of r is: 0.31999999999999995.

Additionally, Python displays the temperature profile graph and the contour plot of the response surface in **Figure 8**:

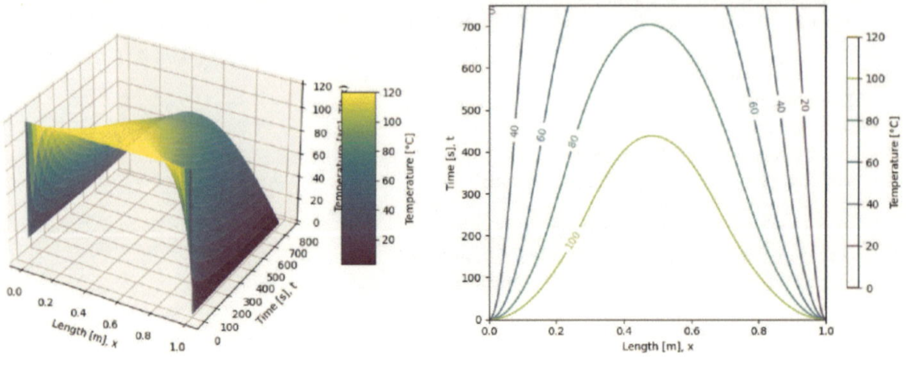

Figure 8.
Temperature distribution in the bar.

2.5.3 Elliptic PDE

To apply the FDM on an elliptic PDE, consider Laplace's eq. [5]:

$$\frac{\partial^2 u}{\partial x^2} + \frac{\partial^2 u}{\partial y^2} = 0 \tag{33}$$

on the domain $x_0 \le x \le x_f$ and $y_0 \le y \le y_f$. The boundary conditions are:

$$u(x_0, y) = q_{x0}(y), \quad u(x_f, y) = q_{xf}(y) \quad u(x, y_0) = q_{y_0}(x), \quad u(x, y_f) = q_{y_f}(x)$$

The domain x is divided into m sections and the domain y is divided into n sections such that $\Delta x = (x_f - x_0)/m$ and $\Delta y = (y_f - y_0)/n$.

Second derivatives are replaced by central difference approximations as follows:

$$\left.\frac{\partial^2 u}{\partial x^2}\right|_{i,j} \cong \frac{1}{(\Delta x)^2}\left(u_{i+1,j} - 2u_{i,j} + u_{i-1,j}\right) \tag{34}$$

$$\left.\frac{\partial^2 u}{\partial y^2}\right|_{i,j} \cong \frac{1}{(\Delta y)^2}\left(u_{i,j+1} - 2u_{i,j} + u_{i,j-1}\right) \tag{35}$$

where $u_{i,j}$ is the value of the function at the point $x_i = x_0 + i(\Delta x)$ and $y_i = y_0 + j(\Delta y)$.

For each interior point (x_i, y_i) with $1 \le i \le m-1$ and $1 \le j \le n-1$, the following finite difference equation is obtained:

$$\frac{1}{(\Delta x)^2}\left(u_{i+1,j} - 2u_{i,j} + u_{i-1,j}\right) + \frac{1}{(\Delta y)^2}\left(u_{i,j+1} - 2u_{i,j} + u_{i,j-1}\right) = 0 \tag{36}$$

where $u_{i,j} = u\left(x_i, y_i\right)$.

Solving $u_{i,j}$ from this equation gives:

$$u_{i,j} = \frac{(\Delta y)^2}{2\left\{(\Delta y)^2 + (\Delta x)^2\right\}}\left(u_{i+1,j} + u_{i-1,j}\right) + \frac{(\Delta x)^2}{2\left\{(\Delta y)^2 + (\Delta x)^2\right\}}\left(u_{i,j+1} + u_{i,j-1}\right) \tag{37}$$

$$u_{0,j} = q_{x0}\left(y_j\right),,, \quad u_{m,j} = q_{xf}\left(y_j\right) \quad u_{i,0} = q_{y_0}(x_i) \quad u_{i,n} = q_{y_f}(x_i)$$

Example 6: A two-dimensional square plate is subjected to prescribed boundary conditions as shown in **Figure 9**. Calculate the temperature distribution on the plate using Laplace's equation for heat transfer.

The equation of the system is

$$\frac{\partial^2 T}{\partial x^2} + \frac{\partial^2 T}{\partial y^2} = 0 \tag{38}$$

on the domain $0 \le x \le 0.2$ and $0 \le y \le 0.2$.

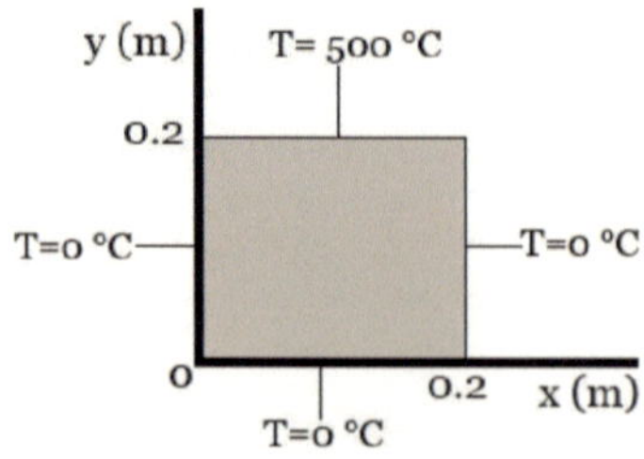

Figure 9.
Two-dimensional square plate.

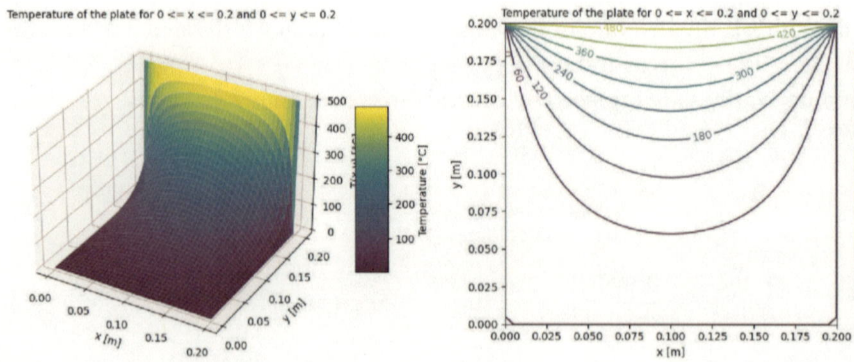

Figure 10.
Temperature distribution and two-dimensional plate surface contour.

The boundary conditions are: $T(0, y) = 0$; $T(0.2, y) = 0$; $T(x, 0) = 0$; $T(x, 0.2) = 500$

To solve this problem the Python file shown below is used. Python returns the response shown in **Figure 10**.

```
# Start: Elliptic PDE
import numpy as np
import matplotlib.pyplot as plt

def elliptic_pde(qx0, qxf, qy0, qyf):
    # Dimensions of the plate in x and y axes
    x0 = 0 # Initial coordinate value on the x-axis
    xf = 0.2 # Final coordinate value on the x-axis
    y0 = 0 # Initial coordinate value on the y-axis
    yf = 0.2 # Final coordinate value on the y-axis

    # Input parameters
    m = 40 # Number of divisions in the x domain
    n = 40 # Number of divisions in the y domain
    critico = 1e-8 # Convergence criterion
    kmax = 10000 # Maximum number of iterations
    delta_x = (xf - x0) / m # Increment in the x-axis
    x = np.linspace(x0, xf, m + 1) # x vector
    delta_y = (yf - y0) / n # Increment in the y-axis
    y = np.linspace(y0, yf, n + 1) # y vector
```

```
# Squared increments
delta_x_squared = delta_x**2
delta_y_squared = delta_y**2
delta_xy_squared = 2 * (delta_x_squared + delta_y_squared)

# Ratios used in the iterative solution
Ratio_x = delta_x_squared / delta_xy_squared
Ratio_y = delta_y_squared / delta_xy_squared

# Initialization of the temperature matrix T
T = np.zeros((n + 1, m + 1))

# Boundary conditions
for k in range(n + 1):
  T[k, 0] = qx0(y[k]) # Left boundary
  T[k, -1] = qxf(y[k]) # Right boundary
for k in range(m + 1):
  T[0, k] = qy0(x[k]) # Bottom boundary
  T[-1, k] = qyf(x[k]) # Top boundary

# Iterative solution using the finite difference method
for k in range(kmax):
  T0 = T.copy()
  for i in range(1, m):
    for j in range(1, n):
      T[j, i] = Ratio_y * (T[j + 1, i] + T[j - 1, i]) + Ratio_x * (T[j, i + 1] + T[j, i - 1])
  # Check for convergence
  if k > 1 and np.max(np.abs(T - T0)) < critico:
    break

# Plot results in 3D
X, Y = np.meshgrid(x, y)
fig = plt.figure()
ax = fig.add_subplot(111, projection='3d')
surf = ax.plot_surface(X, Y, T, cmap='viridis')
# Add color bar to show temperature scale
cbar = fig.colorbar(surf, ax=ax, shrink=0.5, aspect=5)
cbar.ax.set_ylabel('Temperature [°C]', fontsize=10)

# Set axis labels, ticks, and title
ax.set_xlabel('x [m]', fontsize=10)
ax.set_ylabel('y [m]', fontsize=10)
ax.set_zlabel('T(x,y) [°C]', fontsize=10)
ax.set_title('Temperature of the plate for 0 <= x <= 0.2 and 0 <= y <= 0.2', fontsize=10)

# Set axis ticks to increments of 0.05
ax.set_xticks(np.arange(x0, xf + 0.05, 0.05))
ax.set_yticks(np.arange(y0, yf + 0.05, 0.05))

plt.tight_layout() # Adjust margins for better layout
plt.show()

# Plot contour results
fig2, ax2 = plt.subplots()
CS = ax2.contour(X, Y, T, levels=8)
ax2.clabel(CS, inline=True, fontsize=10)

# Add color bar to contour plot
cbar2 = plt.colorbar(CS, ax=ax2, shrink=0.8)
```

```
        cbar2.ax.set_ylabel('Temperature [°C]', fontsize=10)
        # Set axis labels, ticks, and title
        ax2.set_xlabel('x [m]', fontsize=10)
        ax2.set_ylabel('y [m]', fontsize=10)
        ax2.set_title('Temperature of the plate for 0 <= x <= 0.2 and 0 <= y <= 0.2', fontsize=10)
        # Set axis ticks to increments of 0.025
        ax2.set_xticks(np.arange(x0, xf + 0.025, 0.025))
        ax2.set_yticks(np.arange(y0, yf + 0.025, 0.025))

        plt.tight_layout() # Adjust margins for better layout
        plt.show()

        return T, x, y

# Definition of functions for boundary conditions
qx0 = lambda y: 0 # Boundary condition for x, left side
qxf = lambda y: 0 # Boundary condition for x, right side
qy0 = lambda x: 0 # Boundary condition for y, bottom side
qyf = lambda x: 500 # Boundary condition for y, top side

# Call the function to solve the elliptic PDE
T, x, y = elliptic_pde(qx0, qxf, qy0, qyf)
# End
```

2.6 Parametric identification in ordinary differential equations (ODE)

Parametric identification is valuable because it enables the determination of model parameters based on experimental data, ensuring accurate system simulations and predictions. This is crucial in fields such as control systems and engineering. Alternative methods include nonparametric identification, which uses techniques like neural networks without assuming a specific model structure; Bayesian identification, which incorporates prior knowledge and uncertainties for probabilistic parameter estimation; and gray-box modeling, which combines known physical laws with data-driven approaches to identify unknown parameters. These methods provide flexibility, depending on the problem's requirements and constraints [1, 2, 5].

Below is described, through an example, the application of a specific optimization method for the parametric identification of a heat transfer mathematical model. In this example, a Python optimization routine is used.

Example 7: A metal body is heated by an electric resistor (3 W) and is surrounded by a fluid at 23°C. The mathematical model that represents the phenomena that occur during this heating process is eq. (39):

$$mC_p\frac{dT_{(t)}}{dt} = UA\left(T_\infty - T_{(t)}\right) + \epsilon_{body}\sigma A\left({T_\infty}^4 - {T_{(t)}}^4\right) + Q \tag{39}$$

The model parameters are: Initial temperature $(T_0) = 23^{\circ}C$; ambient temperature $(T_\infty) = 23^{\circ}C$; heat source $(Q) = 3$ W; specific heat $(C_p) = 500$ J/kgK; surface area (A) = 1.2 x 10^{-3} m^2; mass (m) = 0.004 kg; Stefan-Boltmann constant (σ) = 5.67 x 10^{-8} W/m^2K^4; in addition, **Table 2** shows the experimental values of the temperature of the metallic body during the heating time. The objective of this problem is to find

t [s]	0	25	50	75	100	125	150	175	200	225	250
T [°C]	296,2	329,5	355,3	374,6	388,8	399,0	406,2	411,3	414,8	417,3	419,0
t [s]	**275**	**300**	**325**	**350**	**375**	**400**	**425**	**450**	**475**	**500**	
T [°C]	420,1	420,9	421,4	421,8	422,1	422,2	422,4	422,4	422,5	422,5	

Table 2.
Experimental data on heating of the metal body.

the values of U and ϵ_{body} from the experimental data and the mathematical model using nonlinear optimization techniques.

To solve this problem, the Python file shown below is used. In this code, curve_fit is used as an optimization routine [4]. This function facilitates the adjustment of a curve to a set of data using the nonlinear least squares method. Entering initial parameter values is optional using curve_fit. The adjustment produces a set of parameters that describe the resulting curve.

```
# Start: Parametric Identification
import numpy as np
import matplotlib.pyplot as plt
from scipy.optimize import curve_fit
from scipy.integrate import odeint

# Experimental data
time = np.array([0, 25, 50, 75, 100, 125, 150, 175, 200, 225, 250, 275, 300, 325, 350, 375, 400, 425, 450, 475,
500])
xdata = np.array([296.15, 329.51, 355.26, 374.6, 388.78, 398.99, 406.22, 411.30, 414.82, 417.27, 418.95,
420.11, 420.90, 421.44, 421.81, 422.07, 422.24, 422.36, 422.44, 422.50, 422.54])

# Inputs
A = 1.2e-3 # Surface area [m^2]
m = 0.004 # Mass [kg]
cp = 500 # Specific heat capacity [J/kgK]
sigma = 5.67e-8 # Stefan-Boltzmann constant [W/m^2K^4]
Q = 3 # Heat input [W]
Ta = 23 + 273.15 # Ambient temperature [K]

# Mathematical model
def model(x, t, U, E):
  return (U * A / (m * cp)) * (Ta - x) + (E * A * sigma / (m * cp)) * (Ta**4 - x**4) + Q / (m * cp)

# Function to integrate the model
def integrate(time, U, E):
  x0 = 296.15 # Initial temperature [K]
  x = odeint(model, x0, time, args=(U, E))
  return x.flatten()

# Fit function for curve_fit
def fit_func(time, U, E):
  return integrate(time, U, E)

# Optimization routine to find the best parameters U and E
params, params_covariance = curve_fit(fit_func, time, xdata, p0=[1e-5, 1e-5])
# Optionally, you can provide an initial guess for the parameters U and E, here: p0= [1e-5, 1e-5].
```

```
# Print the optimized values of U and E
print(f"The optimized value of U is: {params[0]}")
print(f"The optimized value of E is: {params[1]}")

# Plot results

t = np.linspace(0, 500, 100)
x = odeint(model, 296.15, t, args=(params[0], params[1]))
plt.plot(time, xdata, 'ko', label='Experimental data') # Experimental data points
plt.plot(t, x, 'k', linewidth=2, label='Fitted model') # Fitted model curve
plt.ylabel('Temperature, [K]')
plt.xlabel('Time [s]')
plt.legend()
plt.show()
# End
```

Python returns the following response: "The optimal value of U is: 9.965518003635731" and "The optimal value of ϵ_{body} is: 0.9032070092490586". In addition, Python returns **Figure 11** which shows experimental data and the fitted curve.

3. Conclusion

Numerical methods in heat transfer are essential for solving complex problems that cannot be easily addressed using analytical methods. Commonly used methods include the Newton-Raphson Method, Fixed-Point Method, Finite Difference Method (FDM), Euler Method, and Runge-Kutta Method. These techniques offer robust tools for achieving accurate solutions to heat transfer problems in diverse applications, which were exemplified in this chapter.

In the chapter, various numerical methods used in heat transfer are compared, highlighting their differences in terms of accuracy, computational efficiency, and stability. The fixed-point method is simple but can be slow and does not always converge, whereas the Newton-Raphson method is faster and more accurate but relies on a good initial estimate. The finite difference method is efficient for problems with regular geometries but may be less accurate for complex geometries. The Euler and

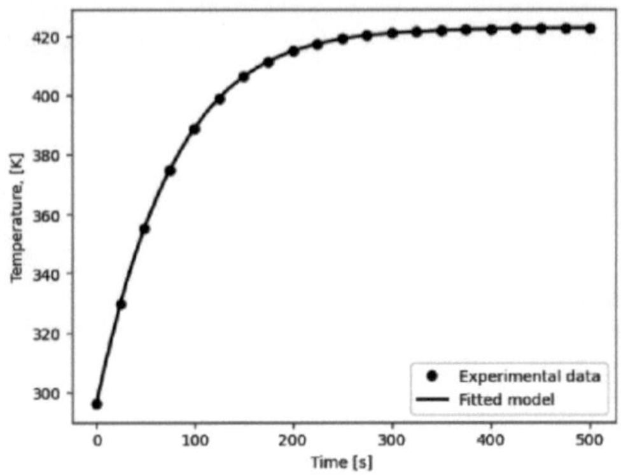

Figure 11.
Curve fitting to experimental data by curve_fit.

Runge-Kutta methods are used to solve ordinary differential equations, with the Runge-Kutta method being more accurate and stable than Euler's method, although more computationally expensive. Finally, Python's solve_ivp function offers a versatile and accurate solution for systems of differential equations, adapting the step size to improve efficiency and accuracy. Stability is crucial in solving heat transfer equations because it ensures that numerical solutions do not diverge and remain significantly meaningful over time.

Nomenclature

A	heat transfer area (m^2)
A_s	surface area of sphere (m^2)
C_p	specific heat (J/kgK)
c_p	specific heat of metal sphere (J/kg K)
$c_{p,c}$	specific heat of fluid in the jacket (J/kgK)
$c_{p,t}$	specific heat of fluid in the tank (J/kgK)
D_1	inner diameter of the pipe (m)
D_2	outer diameter of the pipe (m)
D_3	outer diameter of the insulation (m)
$h_{\infty 1}$	internal convective heat transfer coefficient ($W/m^{2\circ}C$)
$h_{\infty 2}$	external convective heat transfer coefficient ($W/m^{2\circ}C$)
hi	average convection heat transfer coefficient inside the chimney ($W/m^{2\circ}C$)
ho	convection heat transfer coefficient outside the chimney ($W/m^{2\circ}C$)
$k_{concrete}$	thermal conductivity of concrete (W/m°C)
k_{fiber}	thermal conductivity of fiberglass (W/m°C)
k_{steel}	thermal conductivity of steel (W/m°C)
L	pipe length (m)
m	Metal body mass (kg)
m_c	mass of fluid in the jacket (kg)
$\dot{m}_c$	mass flow of fluid in the jacket (kg/s)
m_t	mass of fluid in the tank (kg)
q	heat losses (W)
Q	heat source (W)
$T_{(t)}$	Sphere temperature (K)
$T_{(0)}$ or T_i	initial value of Temperature of the sphere (K)
Ti	average temperature of the hot gases in the chimney (°C)
T_{in}	temperature of the fluid at the jacket inlet (K)
$T_{c(t)}$	temperature of the fluid in the jacket (K)
To	ambient air temperature (°C)
$T_{surroundings}$	Air temperature surrounding the sphere (K)
$T_{t(t)}$	temperature of the fluid in the tank (K)
$T_{\infty 1}$	saturated steam temperature (°C)
$T_{\infty 2}$	environment temperature (°C)
t_f	final processing time (s)
U	overall heat transfer coefficient (W/m^2K)
u	represents the temperature variable (°C or K)
V	volume of the metal sphere (m^3)

Greek letters

α	thermal diffusivity (m^2/s)
ϵ_{body}	emissivity of the metal body (dimensionless)
ϵ_{sphere}	emissivity of the metal sphere (dimensionless)
ϵ_{wall}	emissivity of the outer surface of the wall of the chimney (dimensionless)
ρ	density of the metal sphere (kg/m^3)
σ	Stefan-Boltzmann constant (W/m^2K^4)

Author details

Fabián A. Ortega-Quintana
Universidad de Córdoba, Montería, Colombia

*Address all correspondence to: fortega@correo.unicordoba.edu.co

References

[1] Gupta A. Numerical Methods Using Matlab. California: Apress Berkeley; 2014. 137 p. DOI: 10.1007/9781484201541

[2] Esfandiari R. Numerical Methods for Engineers and Scientists Using Matlab. 2nd ed. Boca Raton: CRC Press; 2017. 493 p. DOI: 10.1201/9781315152417

[3] Pine D. Introduction to Python for Science and Engineering. 2nd ed. Boca Raton: CRC Press; 2024. 444 p. DOI: 10.1201/9781032673950

[4] Çengel Y, Ghajar A. Heat and Mass Transfer. Fundamentals & Applications. 5th ed. New York: McGraw Hill; 2015. 968 p

[5] Yeo Y. Chemical Engineer Computation with Matlab. 2nd ed. Boca Raton: CRC Press; 2020. 847 p. DOI: 10.1201/9781003090601

[6] Chapman S. Matlab® Programming with Applications for Engineers. 1st ed. Canada: CENGAGE Learning™; 2013. 569 p

Chapter 3

Numerical Modeling of Heat and Mass Transfer in the Processes of Oil Deposits Operation in a Deformable Environment

Rasulov Mahir, Jalalov Garib and Sinsoysal Bahaddin

Abstract

In this chapter, a hydrodynamic model that allows taking into account the heat and mass transfer process in a theoretical study of oil field development is proposed. It is known that in solution of a nonlinear equations of the mathematical model describing the filtration process, points "localization", interior discontinuity, etc., that requires an accurate approach at creation numerical methods. In connection, three different finite-difference schemes that can correctly reflect such physical properties are proposed. The first of the proposed finite difference schemes allows for the calculation of the technological parameters of the exploitation using homogeneous schemes that do not take into account the physical properties in the equations, but its application to solving practical problems can lead to certain inaccuracies. To overcome this drawback and make the right decision when solving practical problems, an effective finite difference scheme in the class of generalized functions is proposed. Based on the proposed algorithms, computer tests were carried out on real and laboratory data. This model allows for determining the technological parameters of the exploitation, as well as the actual exploitation life.

Keywords: mass-heat transfer, deformable collector, weak solution, finite differences in a class of generalized functions, auxiliary problem

1. Introduction

During the development of oil reservoirs, the fluids contained in them may acquire a temperature different from the natural temperature of the reservoirs. Temperature changes in productive reservoirs may be caused by various reasons, depending on the nature of the filtration of natural fluids and the type of artificial thermal impact on productive reservoirs during the application of a particular development technology.

Three main groups of reasons leading to non-isothermal filtration conditions can be distinguished [1].

- The action of thermodynamic effects during the movement of reservoir fluids in a porous medium
- barothermal effect (Joule-Thomson and adiabatic expansion effects);
- geothermal gradient;
- effect of phase transformations (during the release of dissolved gas, paraffin precipitation, etc.);
- gravitational effect.

During field development, various combinations and combinations of the above processes arise, which leads to a non-isothermal nature of filtration. Thermal impact on the formation changes the main filtration parameters: fluid viscosity, capillary forces, rheological properties of moving agents, etc. In this case, displacement coefficients, phase permeabilities, etc., change, as a result of which the temperature factor significantly affects both the current filtration characteristics and the final oil recovery. When considering the hydrodynamics of non-isothermal filtration processes, the use of differential equations is no longer sufficient. In this case, a new unknown variable appears – temperature T, and the characteristics of the fluid (its density ρ and viscosity coefficient η) change along with T: $\rho = \rho\ (p, T), \eta = \eta\ (p, T)$. To close the system of equations, it is necessary to additionally involve an equation that determines the change in fluid temperature in time and space. This equation can be obtained by writing down the law of conservation of energy (the first law of thermodynamics) for a reservoir system. But the reservoir rocks and the fluids saturating them have different thermodynamic and rheological properties. Therefore, when writing this law, it is necessary to introduce two temperatures: the temperature of the liquid T and the temperature of the skeleton T_{sk}.

General considerations show that the difference between the temperatures of the liquid and solid phases during filtration should quickly disappear due to the huge heat exchange surface between the fluids and the skeleton, so that the temperatures can be considered equal [1].

Experience in oil field development shows that the complexity of the physical processes that occur during the exploitation of the fields highlights the need for a comprehensive new scientific approach in the creation of technological schemes. The most important direction of these studies is the study of the mechanisms of mass transfer of hydrocarbons and the transition between phases, taking into account changes in pressure and temperature during the filtration of fluid in the formation. It is important to take into account the impact of pressure and temperature changes in the development of hydrocarbon fields on the performance.

Suppose that a circular oil reservoir with a deformed porous medium of height H is exploited by a well with production Q and with radius r_c. Suppose that the permeability in the reservoir changes according to the law $\overline{k}(p) = a_k\left[1 + \alpha_k\ (p - p_s\ )\right]$, where α_k and a_k are positive constants calculated from field or laboratory data. The pressure in wellbore and contour of layer is p_c and p_k, respectively.

Determinate the dynamical distribution of the pressure and temperature functions in the reservoir under these conditions is reduced to found to solution the following system of equations, [1–7][1]

$$\beta^* \frac{\partial p}{\partial t} = \frac{1}{r}\frac{\partial}{\partial r}\left(\frac{k(p)}{\mu} r \frac{\partial p}{\partial r}\right), r_c \le r \le r_k, t > 0 \tag{1}$$

$$C_{pl}\frac{\partial T}{\partial t} = \rho_f C_f \frac{k(p)}{\mu}\frac{\partial p}{\partial r}\left(\frac{\partial T}{\partial r} + \varepsilon \frac{\partial p}{\partial r}\right) + \eta \rho_f C_f \frac{\partial p}{\partial t}, \quad r_c \le r \le r_k, t > 0 \tag{2}$$

$$p(r,0) = p_c, \quad r_c \le r \le r_k \tag{3}$$

$$T(r,0) = T_c, \quad r_c \le r \le r_k \tag{4}$$

$$p(r_k, t) = p_k, t > 0 \tag{5}$$

$$\frac{2\pi k(p)H}{\mu}\left(r\frac{\partial p}{\partial r}\right)\bigg|_{r=r_c} = Q + C\frac{\partial p}{\partial t}\bigg|_{r=r_c}, t > 0 \tag{6}$$

$$T(r_k, t) = T_k, \quad r_c \le r \le r_k. \tag{7}$$

When setting problems (1)–(7), the following physical assumptions are made:

- The well is complete according to its opening degree and character, that is, the well was drilled along the entire height of the oil field, as shown in **Figure 1**,
- Oil filtration obeys Darcy's law and it is non-isothermal,
- The pressure at the bed boundary is constant,

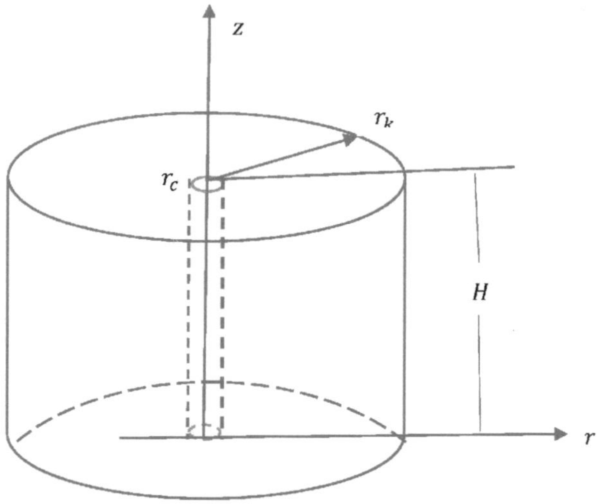

Figure 1.
Model of the layer.

[1] The first equation is the filtration equation obtained using the law of conservation mass equation and Darcy's law, which expresses the filtration rate, and the second equation expresses the energy of the layer the derivation of this is given in Appendix A.

- The initial natural distribution of temperature along the layer is stable,
- The temperature of the fluid and porous medium is the same at any point in the layer,
- Heat transfer in the radial direction of the layer is negligible compared to the convective heat transfer,
- Influence of temperature changes in layer is not taken into account to the parameters defining of porous medium and fluid,
- The variation of the permeability parameter of the layer based on the deformation of the collector is expressed by a well-known empirical formula.

It is clear that it is impossible to find an analytical, that is, exact, solution to this problem. In this case, approximate methods of solution are necessary. One of the most universal approximate methods of solution is the finite difference method. The application of this method and the results obtained are closely related to the geometric structure of the region. In the case under consideration, the fact that the region itself is radial creates certain difficulties in applying the finite differences method. Therefore, we first smooth the region of the problem under consideration by substituting $x = \frac{\overline{r}^2}{4}$, and simultaneously perform a dimensionless transformation in the system of equations. After performing these operations, the question under consideration takes the following form:

$$\frac{\partial \overline{p}}{\partial \tau} = \frac{\partial}{\partial x}\left(x\frac{\overline{k}(\overline{p})}{\overline{\mu}}\frac{\partial \overline{p}}{\partial x}\right), x_c \le x \le x_s, \tau > 0, \quad (8)$$

$$\overline{p}(x, 0) = p_0, x_c \le x \le x_s, \quad (9)$$

$$\frac{4\pi\overline{k}(\overline{p})H}{\overline{\mu}}\left(x\frac{\partial \overline{p}}{\partial x}\right)\Big|_{x=x_c} = \left(\tilde{Q} + \tilde{C}\frac{\partial \overline{p}}{\partial \tau}\right)\Big|_{x=x_c}, \quad (10)$$

$$\overline{p}(x_s, \tau) = p_s, \tau > 0, \quad (11)$$

$$\frac{\partial \overline{T}}{\partial \tau} = \tilde{A}\frac{\overline{k}(\overline{p})}{\mu}x\frac{\partial \overline{p}}{\partial x}\left(\frac{\partial \overline{T}}{\partial x} + \tilde{B}\frac{\partial \overline{p}}{\partial x}\right) + \tilde{D}\frac{\partial \overline{p}}{\partial \tau}, x_c \le x \le x_s, \tau > 0, \quad (12)$$

$$\overline{T}(x, 0) = T_0, \quad (13)$$

$$\overline{T}(x_s, \tau) = T_s. \quad (14)$$

Here,

$$\tau = \frac{k_0}{\mu_0 r_k^2 \beta^*}t, r_0 = \frac{r_c}{r_k}, \; r_s = \frac{r_k}{r_k},$$

$$\tilde{Q} = \frac{\mu_0}{Hk_0 p_k}Q, \tilde{C} = \frac{C}{Hr_k^2\beta^*}, \tilde{A} = \frac{\rho_f C_f p_k \beta^*}{CS_{pl}}, \tilde{B} = \frac{p_k \varepsilon}{T_k}, \tilde{D} = \frac{\eta\rho_f C_f p_k}{C_{pl} T_k}.$$

As it is seen that one of the equations is nonlinear parabolic with respect to the pressure function, and the other is hyperbolic with respect to the temperature function.

It is clear that the localization effect exists in the solution of the nonlinear parabolic type equation that theoretically describes the pressure propagation process in the reservoir [8–10]. This means that there is such a number T_0 and points $x = l(t)$ by $t \geq T_0 \, p(l(t), t) = 0$.

Here $x = l(t)$ denotes the front of distribution of pressure function. It is clear that at the points $x = l(t)$, $\frac{\partial p}{\partial x}$ does not exist. In this case, the points $x = l(t)$ become discontinuity points of the second kind for $\frac{\partial^2 p}{\partial x^2}$, which we will call interior discontinuity points.

As it is note above in processes expressed by such equations, the support of the solution can propagate over a finite distance in a finite time. Mathematically, this means that the solution to the problem has weak discontinuity points of second type, that is, $\operatorname*{supp}_x p(x,t) = \{x_c < x \leq l(t), t > 0\}$, [4, 9]. In order words, the distribution of the pressure has the nature of localization. The localization effect present in the pressure function affects also the dynamic distribution of the temperature field. In other words, the temperature front extends over a finite distance in a finite time. This violates the smoothness of the both pressure and temperature functions. Hence, the classical solutions to the considered problem may not exist.

Before applying the finite difference method to problems (8)–(12), it can be written as follows,

$$\frac{\partial \overline{p}}{\partial \tau} = \frac{\partial w(x,t)}{\partial x},$$

$$\frac{\partial (\overline{T} - \tilde{D}\overline{p})}{\partial \tau} = \tilde{A} w(x,t) \frac{\partial}{\partial x} (\overline{T} + \tilde{B}\overline{p}),$$

$$w(x,t) = -\frac{\overline{k}(\overline{p})}{\mu} x \frac{\partial \overline{p}}{\partial x}.$$

Since $-\frac{\overline{k}(\overline{p})}{\mu} x \frac{\partial \overline{p}}{\partial x}$ from a physical point of view, it represents the flow fluid than the $w(x,t)$ is continuous function. Taking into account the above properties of the both functions $p(x,t)$ and $T(x,t)$, we will introduce the following auxiliary system of equations [11–15]. Let, $D(\bullet) = \partial(\bullet)/\partial x$ and if we apply $D^{-1}(\bullet)$ to both sides of the system of equations, we get

$$\frac{\partial}{\partial \tau} \int_{x_c}^{x} \overline{p}(\xi, \tau) d\xi = \int_{x_c}^{x} w(\xi, \tau) d\xi, \tag{15}$$

$$\frac{\partial}{\partial \tau} \int_{x_c}^{x} (\overline{T}(\xi, \tau) - \tilde{D}\overline{p}(\xi, \tau)) d\xi = \int_{x_c}^{x} \tilde{A} w(\xi, \tau) \frac{\partial}{\partial \xi} (\overline{T}(\xi, \tau) + \tilde{B}\overline{p}(\xi, \tau)) d\xi. \tag{16}$$

System of (15) and (16) we shall say as auxiliary equations as it is seen from system of Eqs. (15) and (16), the functions $p(x,t)$ and $T(x,t)$ outgoing in it can be discontinuous functions, which is an advantage before (8) and (12). Thus, the domain of the functions p and T included in Eqs. (15) and (16) is extended up to class of weak solutions.

2. Finite difference scheme

For the problem under consideration, three different finite difference schemes were constructed:

- Homogeneous finite differences schema,
- Hybrid finite differences schema,
- Finite differences schema in the class of discontinuous functions.

Above, we noted that solutions nonlinear parabolic type equations have some properties that solutions linear equations do not have in differential equations: localization, a finite velocity movement of the front line of the excited region, etc. In the literature, there are schemes in which finite difference schemes are constructed at the nodes of the network without taking into account the features of the equations, and solutions are obtained using standard formulas [16, 17].

2.1 Homogenous implicit scheme for pressure and temperature

Now, we cover the region $[x_c, x_s] \times [0, T)$ by the following grid with a step of h_x and h_τ, as

$$\Omega_{h_x,h_\tau} = \{x_j = x_c + jh_x, j = 0, 1, \dots, n; \tau_k = kh_\tau, k = 0, 1, 2, \dots\}.$$

The problem (8)–(14) at any points (x_j, τ_k) of the grid Ω_{h_x,h_τ} is approximated by the following finite difference scheme as

$$\begin{aligned} P_{j,k+1} = P_{j,k}\left[1 - \gamma\left(R_{j+\frac{1}{2}}(x)K_{j+\frac{1}{2}} + R_{j-\frac{1}{2}}(x)K_{j-\frac{1}{2}}\right)\right] + \gamma R_{j+\frac{1}{2}}(x)K_{j+\frac{1}{2}} P_{j+1,k} \\ +\gamma R_{j-\frac{1}{2}}(x)K_{j-\frac{1}{2}} P_{j-1,k}, \ (j = 1, 2, \dots, n-1, k = 0, 1, 2, \dots) \end{aligned} \tag{17}$$

$$P_{j,0} = 1, (j = 0, 1, \dots, n), \tag{18}$$

$$\gamma_1 \frac{4\pi\bar{k}(P_{0,k})}{\bar{\mu}} (P_{1,k} - P_{0,k}) - \gamma_1\tilde{Q} = \tilde{C}\, P_{0,k+1} - \tilde{C}\, P_{0,k}, \tag{19}$$

$$P_{n,k} = 1, \tau > 0, \tag{20}$$

$$\begin{aligned} T_{j,k+1} = T_{j,k} + \gamma\tilde{A}x_i \frac{\bar{k}(P_{j,k})}{\bar{\mu}} (P_{j+1,k+1} - P_{j,k+1}) \frac{T_{j+1,k} - T_{j-1,k}}{2} + \gamma\tilde{A}Bx_i \frac{\bar{k}(P_{j,k})}{\bar{\mu}} \\ (P_{j+1,k+1} - P_{j,k+1})^2 + \gamma_1\tilde{D}(P_{j,k+1} - P_{j,k}), (j = 1, 2, \dots, n-1; k = 0, 1, 2, \dots,), \end{aligned} \tag{21}$$

$$T_{j,0} = 1, (j = 0, 1, 2, \dots, n), \tag{22}$$

$$T_{0,k} = 1, (k = 0, 1, 2, \dots,) \tag{23}$$

Here $\gamma = \frac{h_\tau}{h_x^2}$ and the network functions $P_{j,k}$ and $T_{j,k}$ are approximate values of functions $\bar{p}(x_j, \tau_k)$ and $T(x_j, \tau_k)$ at an arbitrary points (x_j, τ_k) of the network Ω_{h_x,h_τ}, respectively. As can be seen, (17)–(23) is a classical explicit scheme and has

$O(h_x^2 + h_\tau)$ order of approximation accuracy. It can be easily shown that this scheme is consistent. The for $\gamma \le 0.5$ convergence of this scheme is proven in [18], which also undergoes by computer tests conducted numerous.

2.2 Hybrid finite difference scheme

Since the solution of the nonlinear parabolic type equation has a localization effect, the classical solution of the Eq. (1) may not exist [8–10]. As stated, it is clear that the properties of the order of differentiability of function $\overline{p}(x,t)$ is one unit less than the order of differentiability required of Eq. (1) from the solution. This does not allow us to solve the Eq. (8) by finite difference method. For this reason, account into consideration (15) instead of Eq. (8), we will use the following equation:

$$\frac{\partial}{\partial \tau}\int_{x_c}^{x} \overline{p}(\xi,\tau)d\xi = x\,\frac{\overline{k}(\overline{p})}{\overline{\mu}}\frac{\partial p(x,\tau)}{\partial x} - x_c\,\frac{\overline{k}(\overline{p})}{\overline{\mu}}\frac{\partial \overline{p}(x,\tau)}{\partial x}\bigg|_{x=x_c}, \tag{24}$$

$$\frac{\partial \overline{T}}{\partial \tau} = \tilde{A}\frac{\overline{k}(\overline{p})}{\mu}x\frac{\partial \overline{p}}{\partial x}\left(\frac{\partial \overline{T}}{\partial x} + \tilde{B}\frac{\partial \overline{p}}{\partial x}\right) + \tilde{D}\frac{\partial \overline{p}}{\partial \tau}, x_c \le x \le x_s, \tau > 0, \tag{25}$$

As can be seen, in this case, we use the integro-differential Eq. (24) for the pressure function equivalent to (1). The problem for this system of Eqs. (24) and (25) we will call the first auxiliary problem.

As can be seen, Eq. (8) does not include the second derivative of the function $\overline{p}(x,t)$ with respect to x, which does not exist. Thus, there is no difficulty in applying the finite difference method to Eq. (8).

For Eqs. (24) and (25), the following finite difference scheme is proposed

$$\mathcal{P}_{i,k+1} = \mathcal{P}_{i,k} - \sum_{j=1}^{i-1}(\mathcal{P}_{j,k+1} - \mathcal{P}_{j,k}) + \frac{1}{\overline{\mu}}\frac{h_\tau}{h_x^2}x_i\,k(\mathcal{P}_{i,k})(\mathcal{P}_{i,k} - \mathcal{P}_{i-1,k}) - \frac{x_c\overline{\mu}\,h_\tau}{4\pi \mathrm{H} h_x}\left[\tilde{Q} + \tilde{C}\frac{(\mathcal{P}_{0,k+1} - \mathcal{P}_{0,k})}{h_\tau}\right], \tag{26}$$

$$T_{j,k+1} = T_{j,k} + \gamma\tilde{A}x_i\frac{\overline{k}(\mathcal{P}_{j,k})}{\overline{\mu}}(\mathcal{P}_{j+1,k+1} - \mathcal{P}_{j,k+1})\frac{T_{j+1,k} - T_{j-1,k}}{2} + \gamma\tilde{A}\tilde{B}x_i\frac{\overline{k}(\mathcal{P}_{j,k})}{\overline{\mu}} (\mathcal{P}_{j+1,k+1} - \mathcal{P}_{j,k+1})^2 + \gamma_1\tilde{D}(\mathcal{P}_{j,k+1} - \mathcal{P}_{j,k}), (j = 1, 2, \dots, n-1; k = 0, 1, 2, \dots,). \tag{27}$$

The initial and boundary conditions for system of Eqs. (26) and (27) are

$$\mathcal{P}(x_i, 0) = p_s, i = 0, 1, 2, \dots, n, \tag{28}$$

$$\mathcal{P}(x_n, \tau_k) = p_s, k > 0 \tag{29}$$

$$T_{i,0} = 1, (\, i = 0, 1, 2, \dots, n), \tag{30}$$

$$T_{0,k} = 1, (k = 0, 1, 2, \dots,). \tag{31}$$

As can be seen, (26)–(31) is explicit scheme and also has $O(h_x^2 + h_\tau)$ order of approximation accuracy. It can be easily shown that this scheme is consistent. In addition to Eq. (24) is conservative that shows own advantage. The for $\gamma \le 0.5$ convergence of this scheme that also undergo by computer tests conducted.

2.3 Finite differences schema in a class of generalized function

The above-mentioned localization property of the pressure, that is, a finite velocity movement of the front line of the excited region function, also violates the smoothness of the function expressing the temperature in the layer. In other words, the temperature, as a function of pressure, spreads over a finite distance in a finite time. Under these conditions, we can no longer apply in directly finite difference method to Eq. (2). Therefore, in this section, instead of (2), we will use Eq. (16).

Therefore, in this section, we will construct a difference scheme for the following system of equations

$$\frac{\partial}{\partial\tau}\int_{x_c}^{x}\overline{p}(\xi,\tau)d\xi = x\,\frac{\overline{k}(\overline{p})}{\overline{\mu}}\frac{\partial p(x,\tau)}{\partial x} - x_c\left.\frac{\overline{k}(\overline{p})}{\overline{\mu}}\frac{\partial\overline{p}(x,\tau)}{\partial x}\right|_{x=x_c},$$

$$\frac{\partial}{\partial\tau}\int_{x_c}^{x}(\overline{T}(\xi,\tau)-\tilde{D}\overline{p}(\xi,\tau))d\xi = \tilde{A}\frac{\overline{k}(\overline{p}(\xi,\tau))}{\mu}\xi\frac{\partial\overline{p}(\xi,\tau)}{\partial\xi}\left[\overline{T}(\xi,\tau)+\tilde{B}\overline{p}(\xi,\tau)\right]\Big|_{\xi=x_c}^{x} -$$

$$-\tilde{A}\int_{x_c}^{x}\frac{\partial}{\partial\xi}\left(\frac{\overline{k}(\overline{p}(\xi,\tau))}{\mu}\xi\frac{\partial\overline{p}(\xi,\tau)}{\partial\xi}\right)(\overline{T}(\xi,\tau)+\tilde{B}\overline{p}(\xi,\tau))d\xi,$$

In order to construct a finite differences scheme for (32), we will act as follows. First, let us write this equation in the following form

$$\frac{\partial}{\partial\tau}\sum_{j=1}^{i}\int_{x_{j-1}}^{x_j}\overline{p}(\xi,\tau)d\xi = x\,\frac{\overline{k}(\overline{p})}{\overline{\mu}}\frac{\partial p(x,\tau)}{\partial x} - x_c\left.\frac{\overline{k}(\overline{p})}{\overline{\mu}}\frac{\partial\overline{p}(x,\tau)}{\partial x}\right|_{x=x_c}.$$

After using the rectangular formula, we can write the following finite difference scheme for the last equation

$$\frac{h_x}{h_\tau}\sum_{j=1}^{i}(\mathcal{P}_{j,k+1}-\mathcal{P}_{j,k}) = x_i\,\frac{k(\overline{p}(x_i,t_k))}{\overline{\mu}}\frac{(\mathcal{P}_{i,k}-\mathcal{P}_{i-1,k})}{h_x} -$$

$$\frac{1}{4\pi\mathrm{H}}\left[\tilde{\mathrm{Q}}+\tilde{\mathrm{C}}\frac{(\mathcal{P}_{0,k+1}-\mathcal{P}_{0,k})}{h_\tau}\right].$$

From here, we get

$$\mathcal{P}_{i,k+1} = \mathcal{P}_{i,k} - \sum_{j=1}^{i-1}(\mathcal{P}_{j,k+1}-\mathcal{P}_{j,k}) + \frac{1}{\overline{\mu}}\frac{h_\tau}{h_x^2}x_i\,k(\mathcal{P}_{i,k})(\mathcal{P}_{i,k}-\mathcal{P}_{i-1,k}) - \frac{x_c\overline{\mu}\,h_\tau}{4\pi\mathrm{H}h_x}\left[\tilde{\mathrm{Q}}+\tilde{\mathrm{C}}\frac{(\mathcal{P}_{0,k+1}-\mathcal{P}_{0,k})}{h_\tau}\right]. \quad (32)$$

The finite difference equivalent for Eq. (33) is obtained in a similar manner

$$
\begin{aligned}
\overline{T}(\xi_i,\tau_{k+1}) = {} & \overline{T}(\xi_i,\tau_k) + \tilde{D}\sum_{j=1}^{i}\left[\overline{\mathcal{P}}\left(\xi_j,\tau_{k+1}\right) - \overline{\mathcal{P}}\left(\xi_j,\tau_k\right)\right] - \sum_{j=1}^{i-1}\overline{T}\left(\xi_j,\tau_{k+1}\right) - \overline{T}\left(\xi_j,\tau_k\right) \\
& + \frac{h_\tau}{h_x^2}\tilde{A}\,\frac{\overline{k}\left(\overline{\overline{\mathcal{P}}}(\xi_i,\tau_k)\right)}{\mu}\,\xi_i\left[\overline{\mathcal{P}}(\xi_{i+1},\tau_k) - \overline{\mathcal{P}}(\xi_i,\tau_k)\right]\left[\overline{T}(\xi_i,\tau_k) + \tilde{B}\overline{\mathcal{P}}(\xi_i,\tau_k)\right] \\
& - \tilde{A}\frac{h_\tau}{h_x}\frac{\mu}{4\pi H}\left(\tilde{Q} + \tilde{C}\frac{\overline{\mathcal{P}}(\xi_i,\tau_{k+1}) - \overline{\mathcal{P}}(\xi_i,\tau_k)}{h_\tau}\right)\Bigg|_{x=x_c}\left[\overline{T}(\xi_i,\tau_k) + \tilde{B}\overline{\mathcal{P}}(\xi_i,\tau_k)\right]\Big|_{\xi=x_c} \\
& - \frac{\tilde{A}h_\tau}{\mu h_x^2}\sum_{j=1}^{i}\left\{\overline{k}\left(\overline{\mathcal{P}}\left(\xi_{j+1},\tau_k\right)\right)\xi_{j+1}\left(\overline{\mathcal{P}}\left(\xi_{j+2},\tau_k\right) - \overline{\mathcal{P}}\left(\xi_{j+1},\tau_k\right)\right)\right. \\
& \left. - \overline{k}\left(\overline{\mathcal{P}}\left(\xi_j,\tau_k\right)\right)\xi_j\left(\overline{\mathcal{P}}\left(\xi_{j+1},\tau_k\right) - \overline{\mathcal{P}}\left(\xi_j,\tau_k\right)\right)\right\} \\
& \left(\overline{T}\left(\xi_j,\tau_k\right) + \tilde{B}\overline{\mathcal{P}}\left(\xi_j,\tau_k\right)\right)\ (i = 1,2,\ \dots,n; k = 0,1,2,\ \dots,).
\end{aligned}
\tag{33}
$$

As in the previous sections, the initial and boundary conditions for the system of Eqs. (32)–(33) have the form (28)–(31).

Schemes (32) and (33) are explicit schemes and also have $O\left(h_x^2 + h_\tau\right)$ order of approximation accuracy. It can be easily shown that this scheme is consistent. In addition to Eqs. (32) and (33) are conservative and show own advantages. The for $\gamma \leq 0.5$ convergence of this scheme is also undergone by computer tests conducted.

3. Analysis of calculation results of the computer calculations

Computer tests were carried out based on algorithms (17)–(23) and by given physical parameters, assuming that the absolute permeability of the formation obeys the law $\overline{k}(p) = a_k\left[1 + \alpha_k\ (p - p_s)\right]$ [19].

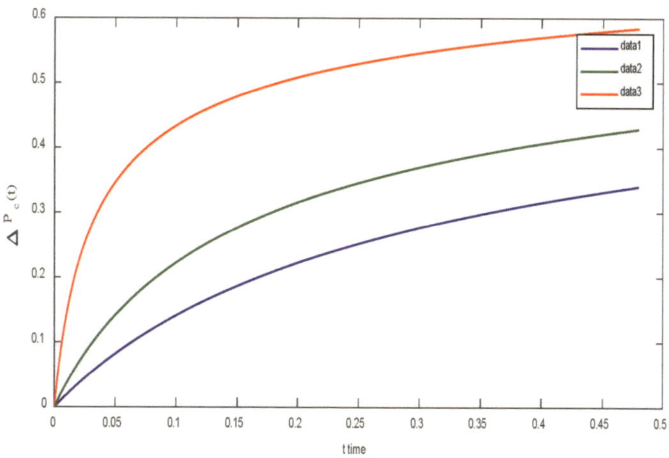

Figure 2.
Time dependence graphs of the dynamic viscosity of the wellbore pressure drop at the specified value of $\mu = 0.5 \cdot 10^{-4} mPa \cdot s$, *and at different values of permeability; (1)* $k = 0.005 \cdot 10^{-12}\ m^2$, (2) $k = 0.01 \cdot 10^{-12}\ m^2$, *(3)* $k = 0.05 \cdot 10^{-12}\ m^2$.

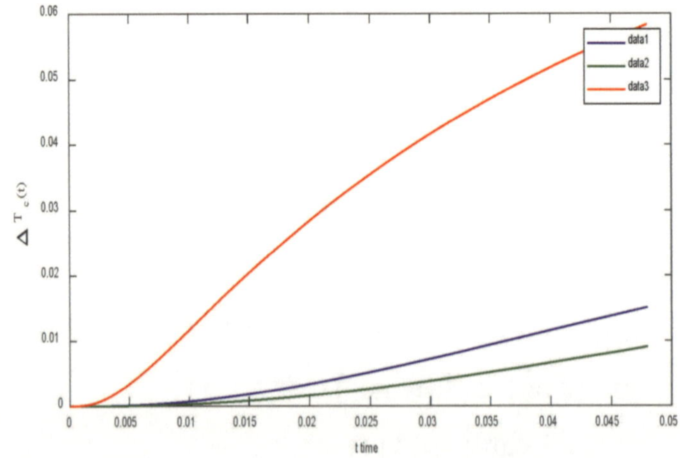

Figure 3.
Graphs of time dependence at different values of viscosity at the specified value of permeability $k = 0.05 \cdot 10^{-12}\ m^2$, *and at different values of viscosity: (1)* $\mu = 1 \cdot 10^{-10} mPa \cdot s$, *(2)* $\mu = 0.5 \cdot 10^{-10} mPa \cdot s$, *(3)* $\mu = 0.7 \cdot 10^{-10} mPa \cdot s$.

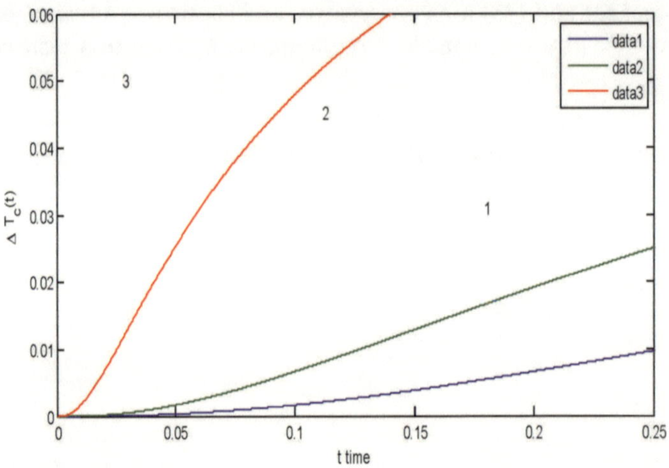

Figure 4.
Graphs of time dependence of the temperature drop at the output of the layer at the specified of value $\mu = 0.7 \cdot 10^{-4} mPa \cdot s$, *and at different values of permeability: (1)* $k = 0.01 \cdot 10^{-12}\ m^2$, *(2)* $k = 0.005 \cdot 10^{-12}\ m^2$ *and (3)* $k = 0.05 \cdot 10^{-12} m^2$.

Under computer tests using the proposed algorithms, only the influence of various parameters on the change in bottomhole pressure and temperature with respect to time was studied. However, the physical interpretation of the results obtained is briefly given in the corresponding captions of figures.

Figures 2–11 showed the time dependence graphs of the thermophysical and filtration-capacity parameters of the fluid temperature at the well bore. It is clear that the change in pressure in the oil reservoir occurs only due to the well, which in turn leads to a change of temperature in the reservoir. **Figures 2** and **3** showed the time dependence graphs of the pressure and the fluid temperature at the wellbore,

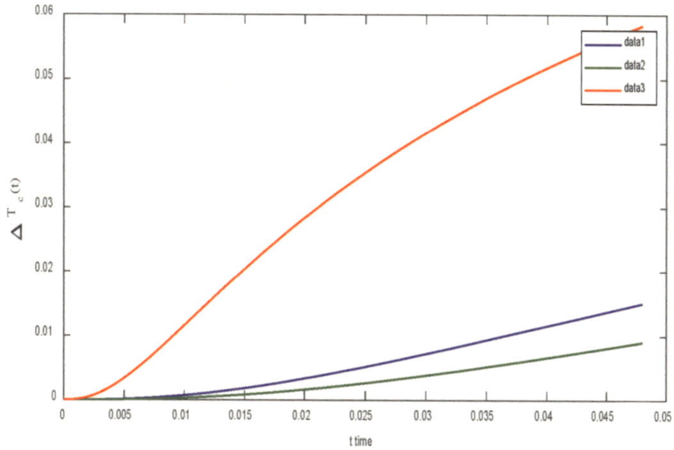

Figure 5.
Graphs of the time dependence of the temperature drop at the output of well bore for a given value $k = 0.05 \cdot 10^{-12} m^2$ *and for different viscosity values: (1)* $\mu = 1 \cdot 10^{-10}$ *mPa•s, (2)* $\mu = 0.5 \cdot 10^{-10} mPa \cdot s$, (3) $\mu = 7 \cdot 10^{-10} mPa \cdot s$.

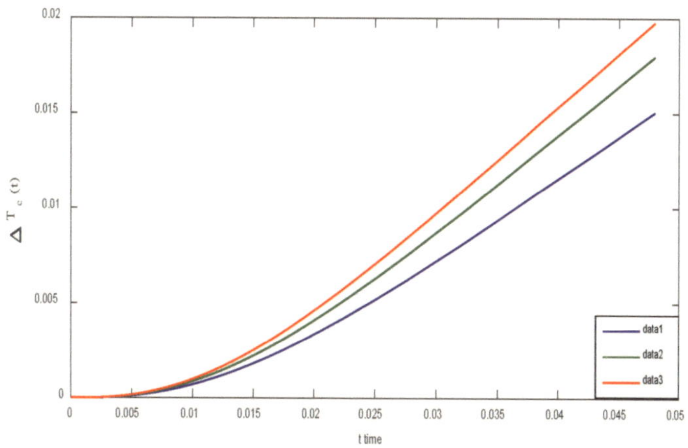

Figure 6.
Graphs of time dependence of the temperature drop at the outlet of the layer at the given value of the Joule-Thomson coefficient $\varepsilon = 0.4 \frac{^\circ C}{MPa}$ *and at different values of the specific heat capacity of the liquid: (1)* $C_{fluid} = 1500 \frac{J}{kg \cdot ^\circ C}$, (2) $C_{fluid} = 1850 \frac{J}{kg \cdot ^\circ C}$, (3) $C_{fluid} = 2100 \frac{J}{kg \cdot ^\circ C}$

respectively. These results show that the change in pressure in the well does not depend on the thermobaric parameters but only on the parameters of the filtration capacity (**Table 1**).

Temperature anomaly is observed when changes in the value of the Joule-Thomson` coefficient more dramatically affect the distribution of the temperature field over time (**Figure 6**). In the variant under consideration, with an increase in the value of the parameter ε, the difference between the temperature drops is about 70–75% compared to approximate $0.1 \frac{^\circ C}{MPa}$.

As can be seen from the graph of the dependence of the temperature drop at the outlet of the formation on the deformation coefficient (**Figure 8**), the deformation

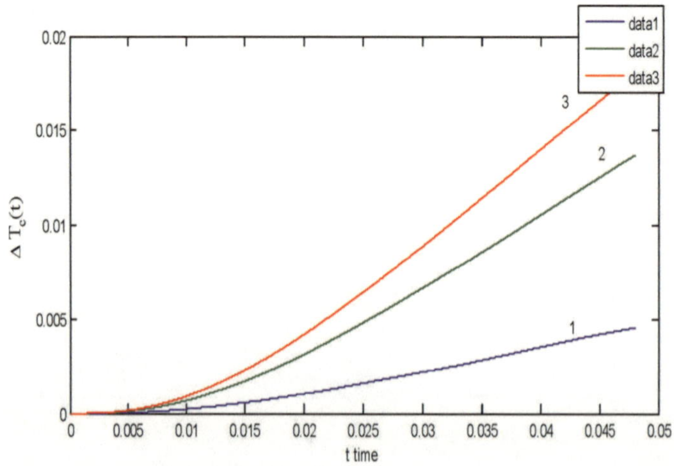

Figure 7.
Graphs of the dependence of the temperature drop at the output of the formation at the given value of heat capacity of the liquid $C_{fluid} = 1880\frac{J}{kg \cdot {}^{\circ}C}$ *and at the different values of the Joule-Thomson effect: (1)* $\varepsilon = 0.1\frac{{}^{\circ}C}{MPa}$, *(2)* $\varepsilon = 0.3\ \frac{{}^{\circ}C}{MPa}$, *(3)* $\varepsilon = 0.4\frac{{}^{\circ}C}{MPa}$

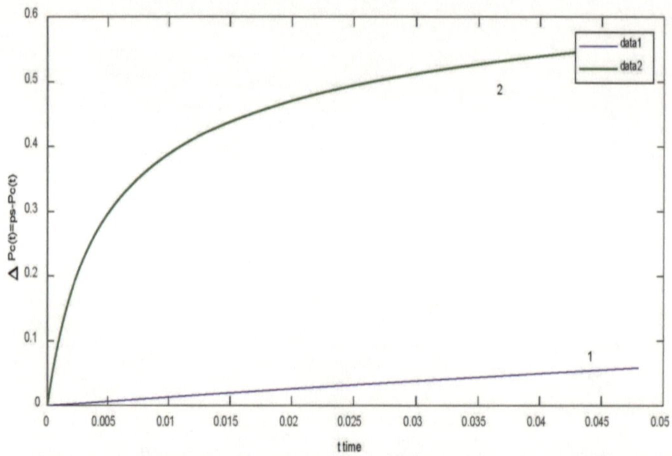

Figure 8.
Dependence of the pressure drop at the outlet of the formation on the well deformation factor: (1) the case $\alpha_k = 0$; *(2) the case* $\alpha_k = 1$.

process in the rock during the development of the formation has an insignificant effect on the temperature anomaly.

Using the algorithm written for system of Eqs. (32) and (33) within initial and boundary conditions of (28) and (31), carry out computer test and count up the distribution of the pressure function along the layer at a given value of time has been analyzed using and the obtained result is shown in **Figure 11**.

As can be seen from the figure, the propagation speed of the pressure function in the scheme written in the class of generalized functions by comparison is closer to the physical propagation speed (**Figure 12**).

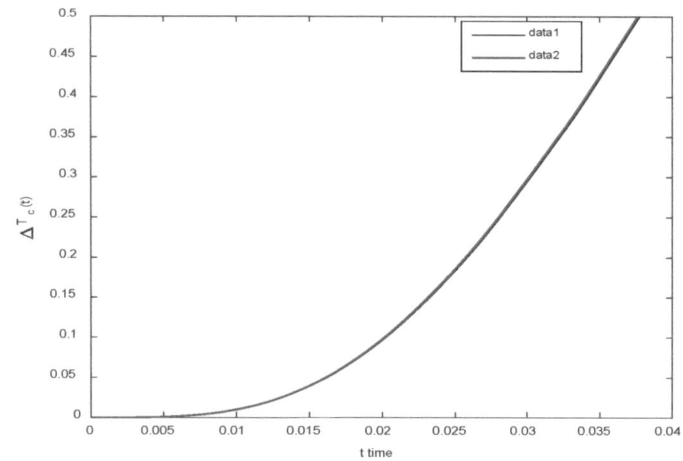

Figure 9.
Dependence of the temperature drop at the outlet of the layer on the deformation factor: (1) the case $\alpha_k = 0$, (2) the case $\alpha_k = 1$.

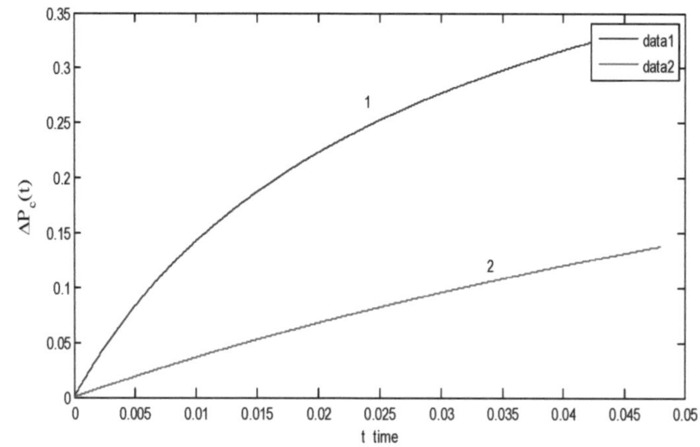

Figure 10.
Dependence of the pressure drop at the outlet of the formation on the coefficient of impact of the wellbore volume: (1) the case $C = 0$; (2) the case $C = 1$.

Some of the materials presented in this work have been partially published in the works [20, 21].

4. Conclusions

- A hydro-thermodynamic model of the process of heat and mass transfer formed in the conditions of nonstationary filtration of a liquid in a porous medium layer with a deformed collector is proposed and algorithms for numerical solutions are developed. The effect of the parameters included in the model to exploiatation process has been studied.

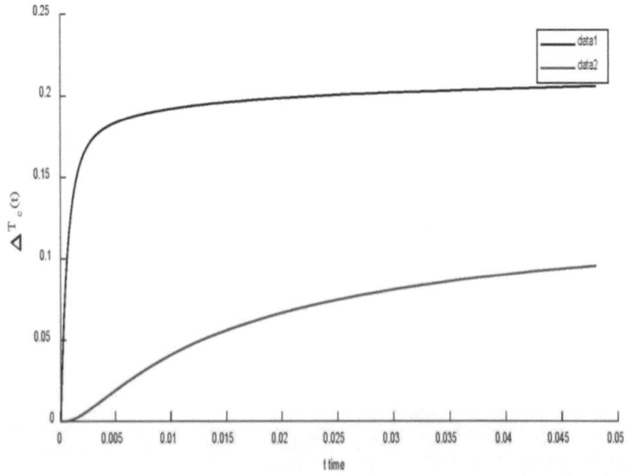

Figure 11.
Dependence of the temperature drop at the outlet of the formation on the coefficient of impact of the wellbore volume: (1) the case C = 0; (2) the case C = 1.

Physical quantities	Numeric values
β^*	$1.0 \cdot 10^{-4}$
k	$0.05 \cdot 10^{-12}$
μ	$7.0 \cdot 10^{-10}$
r_c	0.1
r_k	$0.2 \cdot 10^{2}$
C_{pl}	1800
C_{sr}	$1.8 \cdot 10^{3}$
C_f	$1.5 \cdot 10^{3}$
ρ_f	$8.5 \cdot 10^{2}$
m	0.2
ε	0.4
p_k	40.2
p_c	40.2
T_k	65
H	15
Q	$1.157 \cdot 10^{-4}$
$a_k = A + Bp_k - C\,p_{qor}$	$A = 0.832$, $B = 0.8147 \cdot 10^{-2}$, C = 0.0023
p_{qor}	96.6
η	$1,5 \cdot 10^{-2}$

Table 1.
Values of physical parameters utilised in computer tests.

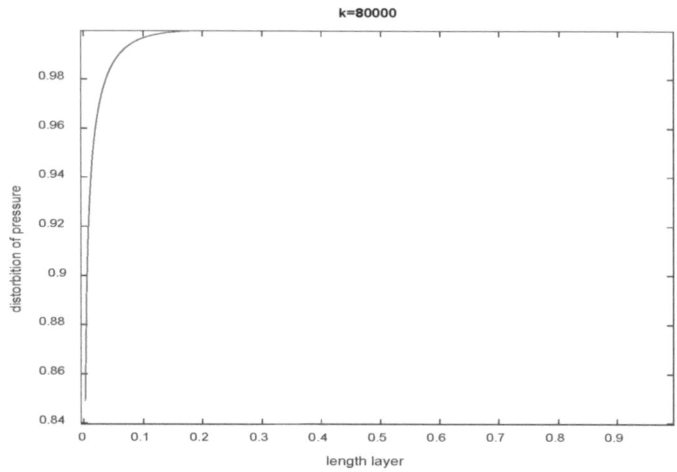

Figure 12
Distribution of the pressure function along the layer at a given value of time

- Based on the results of numerical calculations, it was found that the thermophysical and filtration-capacity parameters have different effects on the distribution of pressure and temperature areas in the formation, which are important to consider in solving practical problems related to field development, including well thermometry.

- The proposed algorithms correctly reflect the dynamic distribution of pressure and temperature functions in the considered problems in principle, which allows for determining the technological parameters accurately. However, the results obtained using a homogeneous difference scheme show that the speeds of the numerical (mathematical) fronts of the pressure and temperature functions are greater than the speed of the physical front (CFL condition), which means that the time of exploiatation of layer ends faster.

- A hybrid finite difference scheme, the results obtained show that the speed of the numerical front of the pressure distribution is less rather than the speed of the numerical front of pressure propagation in the previous scheme proposed. This change, in turn, reduces the motion temperature front.

- Finally, the last third finite difference scheme, that is, the finite difference scheme proposed in the class of generalized functions, shows that the speeds of movement of the fronts of the pressure and temperature functions are closer to the natural physical speed of propagation of movement. This confirms that the time exploitation of a layer is longer and ensures that it can bring high profits.

Nomenclature

β^*	elastic capacity coefficient of the reservoir $\left[\frac{1}{\text{MPa}}\right]$
k	permeability coefficient $[\mu\text{m}^2]$
μ	dynamic viscosity [mPa•s]
r_c	radius of the well [m]

r_k	radius of the layer [m]
t	time variable [s]
C_{pl}	volumetric heat capacity of the reservoir $\left[\frac{J}{m^3 \cdot K}\right]$
C_{sr}	specific heat capacity of the rock $\left[\frac{J}{kg \cdot K}\right]$
C_f	specific heat capacity of the fluid $\left[\frac{J}{kg \cdot K}\right]$
ρ_f	density $\left[\frac{kg}{m^3}\right]$
m	porosity
ε	Joule-Thomson coefficient $\left[\frac{K}{MPa}\right]$
η	coefficient of adiabatic expansion $\left[\frac{K}{MPa}\right]$
p_k	reservoir pressure [MPa]
T_k	reservoir temperature $[K]$
H	reservoir thickness $[m]$
Q	well debit $\left[\frac{m^3}{s}\right]$
C	borehole influence coefficient $\left[\frac{m^3}{MPa}\right]$

A. Derivation of the energy equation for a single-phase filtration flow

The energy balance equation in formations taking into account various factors was considered by Chekalyuk [1].

For a one-dimensional fluid flow in the direction of the x-axis in a cylindrical element of length dx and cross-sectional area ω, the relationship expressing the energy balance for the fluid-porous medium system has the form [1]:

$$d(U+M) = \delta A^{(\varepsilon)} + \delta Q^{(\varepsilon)}, \tag{A1}$$

where U is the internal energy of a certain volume of fluid and skeleton, M is the energy of the fluid in the field of potential forces, $\delta Q^{(\varepsilon)}$ is the amount of heat received by the volume under consideration from the outside; $\delta A^{(\varepsilon)}$ is the work done on it by external forces, written as

$$d(U+M) = \frac{\partial}{\partial t}[m\rho u + (1-m)\rho_{ck}u_{ck} + m\rho gz]\omega dxdt \tag{A2}$$

where m is the porosity; ρ_{sk} is the density of rock grains; z is the elevation mark; u and u_{sk} are the internal energy of a unit mass of liquid and skeleton, respectively.

The heat flow vector in this case can be expressed as

$$q = \rho uw - \lambda \frac{\partial T}{\partial x} \tag{A3}$$

Heat enters through sections x and leaves through section $x + dx$. Therefore

$$\delta Q^{(\varepsilon)} = -\frac{\partial q}{\partial x}\, dx\omega dt = \frac{\partial}{\partial x}(\lambda\, \partial T - \rho uw)\omega dxdt. \tag{A4}$$

The work of external forces $\delta A^{(\varepsilon)}$ performed during movement from section x to position $x + dx$, consists of the work of the pressure forces $\delta A_p^{(\varepsilon)}$ and the force of gravity $\delta A_g^{(\varepsilon)}$, which are respectively equal to:

$$\delta A_p^{(\varepsilon)} = [p(x)w(x) - p(x+dx)w(x+dx)]\omega dt = -\frac{\partial}{\partial x}(pw)\, dx\omega dt;$$

$$\delta A_g^{(\varepsilon)} = [\rho(x)gz(x)w(x) - \rho(x+dx)gz(x+dx)w(x+dx)]\omega dt = -\frac{\partial}{\partial x}(\rho gzw)dx\omega dt.$$

From here we find:

$$\delta A^{(\varepsilon)} = -\frac{\partial}{\partial x}(\rho w + \rho gzw)\omega dxdt. \quad \text{(A5)}$$

Substituting (A2), (A4) and (A5) into equality (A1) and canceling out $\omega dxdt$, we obtain

$$\frac{\partial}{\partial t}[m\rho u + (1-m)\rho_{sk}u_{sk} + m\rho gz] = \frac{\partial}{\partial x}\left(\lambda\frac{\partial T}{\partial x} - \rho uw - pw - \rho gzw\right). \quad \text{(A6)}$$

Using the continuity equation for a one-dimensional flow in the form

$$\frac{\partial(m\rho)}{\partial t} + \frac{\partial(\rho w)}{\partial x} = 0,$$

Eq. (A6) is transformed as follows:

$$\frac{\partial(m\rho u)}{\partial t} + \frac{\partial(\rho uw)}{\partial x} = u\left(\frac{\partial(m\rho)}{\partial t} + \frac{\partial(\rho w)}{\partial x}\right) + m\rho\frac{\partial u}{\partial t} + \rho w\frac{\partial u}{\partial x} = \rho\left(m\frac{\partial u}{\partial t} + w\frac{\partial u}{\partial x}\right);$$

$$\frac{\partial(pw)}{\partial x} = \frac{\partial}{\partial x}\left(\frac{p}{\rho}\rho w\right) = \frac{p}{\rho}\frac{\partial}{\partial x}(\rho w) + \rho w\frac{\partial}{\partial x}\left(\frac{p}{\rho}\right) = \frac{p}{\rho}\frac{\partial(m\rho)}{\partial t} + \rho w\frac{\partial}{\partial x}\left(\frac{p}{\rho}\right);$$

$$\frac{\partial}{\partial t}(m\rho gz) + \frac{\partial}{\partial x}(\rho gzw) = \rho gw\frac{\partial z}{\partial x}.$$

In addition, the last equality takes into account the condition of stationarity and homogeneity of the gravitational field: $\frac{\partial(gz)}{\partial t} = 0, g = const.$

After these transformations, Eq. (A6) takes the form

$$m\rho\frac{\partial u}{\partial t} + \frac{\partial}{\partial t}[(1-m)\rho_{sk}\, u_{sk}] + \rho gw\frac{\partial z}{\partial x} + \rho w\frac{\partial}{\partial x}\left(u + \frac{p}{\rho}\right) - \frac{p}{\rho}\frac{\partial(m\rho)}{\partial t} = \frac{\partial}{\partial x}\left(\lambda\frac{\partial T}{\partial x}\right) \quad \text{(A7)}$$

Taking into account that

$$m\rho\frac{\partial u}{\partial t} - \frac{p}{\rho}\frac{\partial(m\rho)}{\partial t} = m\rho\frac{\partial}{\partial t}\left(u + \frac{p}{\rho}\right) - \frac{\partial(mp)}{\partial t}$$

enthalpy is introduced

$$i = u + \frac{p}{\rho} \tag{A8}$$

and the internal energy of a unit volume of the solid skeleton of the rock is expressed proportionally to its temperature

$$(1 - m)\rho_{sk}\, u_{sk} = c_{sk} T, \tag{A9}$$

where c_{sk} – volumetric heat capacity of the skeleton [J/(m^3 K)], and (A7) is represented as

$$m\rho\,\frac{\partial i}{\partial t} + c_{sk}\,\frac{\partial T}{\partial t} + \rho g w\,\frac{\partial z}{\partial x} + \rho w\,\frac{\partial i}{\partial x} - \frac{\partial (mp)}{\partial t} = \frac{\partial}{\partial x}\left(\lambda\,\frac{\partial T}{\partial x}\right). \tag{A10}$$

The total differential of enthalpy in variables p and T is written in the form

$$di = c_p(dT + \varepsilon_i dp), \tag{A11}$$

where c_p is the isobaric heat capacity of the fluid $\left(c_p = \frac{\partial i}{\partial T}\right)$; ε_i is the Joule-Thomson coefficient.

From equality (A11) we find:

$$\frac{\partial i}{\partial t} = c_p\left(\frac{\partial T}{\partial t} + \varepsilon_i\,\frac{\partial p}{\partial t}\right); \frac{\partial i}{\partial x} = c_p\left(\frac{\partial T}{\partial x} + \varepsilon_i\,\frac{\partial p}{\partial x}\right).$$

Substituting these expressions into (A10), we obtain, after transformations, the desired energy equation for a one-dimensional filtration flow:

$$c_p\,\frac{\partial T}{\partial t} + \rho w c_p\left(\frac{\partial T}{\partial x} + \varepsilon_i\,\frac{\partial p}{\partial x} + \frac{g}{c_p}\,\frac{\partial z}{\partial x}\right) - m\rho c_p\,\eta_s\,\frac{\partial p}{\partial t} = \frac{\partial}{\partial x}\left(\lambda\,\frac{\partial T}{\partial x}\right) \tag{A12}$$

Here, it is taken into account that the porous medium is non-deformable, that is, $\frac{\partial m}{\partial t} = 0$ and the following notations are introduced: $c_\pi = m\rho c_p + c_{sk}$— heat capacity of the saturated porous medium; $\eta_s = \frac{1}{\rho c_p} - \varepsilon_i$—adiabatic cooling coefficient.

Eq. (A12) expresses the fact that the change in temperature of the skeleton-fluid system $\left(c_\pi \frac{\partial T}{\partial t}\right)$ is determined by the following factors:

1. convective heat transfer $\rho w c_p \frac{\partial T}{\partial x}$;

2. throttling effect $\rho w g \frac{\partial z}{\partial x}$

3. the transformation of the potential energy of the gravity field into thermal $\rho w g\, \frac{\partial z}{\partial x}$;

4. the effect of adiabatic cooling $\left(-m\rho c_p \eta_s \frac{\partial p}{\partial t}\right)$;

5. thermal conductivity $\frac{\partial}{\partial x}\left(\lambda\, \frac{\partial T}{\partial x}\right)$.

Eq. (A11) can be generalized to the case of three-dimensional flow:

$$c_p \frac{\partial T}{\partial t} + \rho c_p \vec{w} \left(\nabla T + \varepsilon_i \nabla p + \frac{\vec{g}}{c_p} \right) - m \rho c_p \, \eta_s \frac{\partial p}{\partial t} = div\,(\lambda \nabla T).$$

Author details

Rasulov Mahir[1]*, Jalalov Garib[1] and Sinsoysal Bahaddin[2]

1 Ministry of Science and Education of Azerbaijan, Institute of Oil and Gas, Baku, Azerbaijan

2 Department of Computer Engineering, Faculty of Engineering and Architecture, Istanbul Beykent University, Istanbul, Turkey

*Address all correspondence to: mresulov@gmail.com

References

[1] Çekalyuk EV. Thermodynamics of Oil Reservoir. Moscow: Nedra; 1965. 238 p

[2] Filippov AI, Akhmetova OV. Temperature Field in the Reservoir and the Well. Ufa: Gilem. Ufa; 2011. 329 p (in Russian)

[3] Karachinsky VE. Methods of Geothermodynamics of Gas and Oil Deposits. Moscow: Nedra; 1975. 168 p (in Russian)

[4] Parshin AV. Study of unsteady temperature fields in oil and gas reservoirs as applied to well thermometry [abstract of the dissertation for the degree of candidate of technical sciences]. Ufa; 2012. 23 p (in Russian)

[5] Ramazanov Sh. A. Theoretical foundations of thermohydrodynamic methods for studying oil reservoirs [dissertation for the degree of doctor of technical sciences]. Ufa; 2004. 269 p (in Russian)

[6] Vakhitov GG, Gattenberger Yu P, Lutkov VA. Geothermal Methods of Control over the Development of Oil Fields. Moscow: Nedra; 1984. 240 p (in Russian)

[7] Valiullin RA, Ramazanov Sh A, Sharafutdinov RF. Thermometry of Reservoirs with Multiphase Flows. Ukhta, Ufa: Publishinghouse Bashk; 1998. 115 p (in Russian)

[8] Barenblatt GI, Vishik MI. On the finite rate of propagation of the perturbations in problems of non-stationary filtration of a fluid and a gas. Applied Mathematics and Mechanics. 1956;**20**:411-417

[9] Kalashnikov AS. On a non-linear equation arising in the theory of non-stationary filtration. Proceedings of the Seminar named after I. G. Petrovsky. 1978;**4**:137-146

[10] Kershner R. The behaviour of the solutions of degenerate quasi-linear parabolic equations as $t \to \infty$. Acta Scientiarum Mathematicarum Hungarica. 1979;**34**:157-163

[11] Rasulov M.A. On a method of solving the Cauchy problem for a first-order nonlinear equation of hyperbolic type with a smooth equation. Soviet Mathematics - Doklady, USSR. 1991;**316** (4):777-781. (Abasov MT)

[12] Rasulov MA. To the theory filtration of three-phase mixes in view of them compressible. Soviet Mathematics - Doklady, USSR. 1992;**325**(1):130-133 (Abasov MT)

[13] Rasulov MA. An effective algorithm for Euler systems in a class of discontinuous functions. Mathematical and a Computational Applications;**4**(2): 185-189. USA, (Erhan Coskun)

[14] Mahir R. The finite differences scheme for the first order system of nonlinear differential equations in a class of discontinuous functions. Applied Mathematics and Computation (USA). 2004;**154**:671-681. DOI: 10.1016/S0096-3003(03)00742-2

[15] Rasulov M, Sinsoysal B. Finite Differences Method for Solving of First Order Hyperbolic Type Equations with Non Convex State Function Flow in a Class of Discontinuous Functions. Difference Equations and Applications. USA: Hindawi Publishing Corparation; 2006

[16] Richmayer RD, Morton KW. Difference Methods for Boundary

Problems. New York: Wiley, Int.; 1967. 418 p

[17] Samarskii AA. Theory of Finite Difference Schemes. Moscow: Nedra; 1977. 56 p

[18] Rasulov MA. A numerical method of solution of a nonlinear equation of hyperbolic type of the first order. Differential Equations, USSR. 1992; **28**(7):1254-1261. (Ragimova TA)

[19] Jalalov GI, Ibrahimov TM, Aliyev AA, Gorshkova Ye V. Modeling and investigation of filtration processes in deep oil and gas fields. Elm ve Takhsil. Baku. 2018. 382 p. (in Russian)

[20] Jalalov QI, Rasulov MA, Mirzoyeva DR. Numerical modelling of thermohydrodynamic prosesses in fractured oil reservoir, ANAS transactions. Earth Science. 1/2023: 42-48. DOI: 10.33677/ggianas20230100092

[21] Mahir R, Garib J. Numerical method for studying the process of mass - Heat transfer in deformable layers in a class of discontinuous functions. SOCAR Proceedings. 2023;**4**:72-75. DOI: 10.5510/OGP20230400917

Chapter 4

Modern Dimensional Analysis Involved in Heat Transfer Problems

Ioan Száva, Teofil-Florin Gălățanu and Ildikó-Renáta Száva

Abstract

This chapter introduces a simplified and efficient approach to dimensional analysis, namely the so-called modern dimensional analysis (MDA). MDA is a simple, safe and user-friendly version of the dimensional methods that are currently widely applicable to several engineering problems, such as heat transfer, load-bearing capacity, as well as fire protection of buildings. MDA provides a unique and very simple way to obtain the complete Model Law (ML), which governs the prototype-model correlation. The inferred ML, based on the results of experimental investigations performed exclusively on the attached reduced-scale model, will predict the behaviors of the full-scale prototype. Through its facilities, MDA makes sure to attach the most suitable and flexible model to the prototype and thereby: to obtain in the best manner the required information regarding the given prototype. In this chapter, the currently used dimensional methods are described one by one, along with their advantages and limitations; the MDA principle, with its indisputable advantages; its main stages; useful applications of MDA in the field of heat transfer, as well as a series of conclusions of the authors based on more than 25 years of experience in the application of this very efficient and user-friend method. Section 2 describes the advantages and limitations of the geometric analogy, similarity and classical dimensional analysis, while Section 3: MDA with its main stages and advantages compared to the rest of the classical methods. By comparing them, the reader can confirm or deny the correctness of the authors' claim regarding the remarkable efficiency of MDA.

Keywords: heat transfer, dimensional analysis, prototype-model correlation, fire protection, load-bearing capacity, experimental validation

1. Introduction

In-depth knowledge of the heat transfer phenomenon is important not only from a theoretical point of view but also from a practical point of view, such as the fire resistance of the buildings. Fire resistance represents the time in which the involved structural element maintains its intended function during the action of fire. Starting from this aspect, engineers looked for different ways of approaching the heat transfer in the resistance elements of the structures, in order to predict the propagation of heat in them as accurately, simply and safely as possible.

In the present case, the focus is only on the heat transfer in the structural elements, without touching the theoretical elements specific to the heat transfer. It is well-known that as the temperature rises, the load-bearing capacity of these structural elements decreases significantly, which will lead to their premature deterioration, before the people and material goods in these buildings can be saved.

The main factors on which fire resistance depends must be specified, such as they are: the load-bearing capacity or mechanical strength (the ability to maintain the initial structural integrity of the building during a real or simulated fire), the integrity of involved materials, or their tightness (their ability to stop both the heat and flames through the involved structure subjected to fire), thermal insulation quality/performance of the used/involved materials (their capacity to limit as much as possible the growing of the temperature on the unexposed side of the structural element).

Subsequently, based on this information, the most effective fire protection solutions for buildings can be developed. In this field, but not only, nowadays, it has become very widespread to replace experimental investigations on real parts (prototypes) with those carried out on parts attached to them, made on a convenient scale, called models.

There is no reference to a specific way of selecting fire protection solutions with the help of MDA, but only to the fact that, based on the results of heat propagation along the structural elements, MDA helps to obtain a precise (predicted) map of the thermal fields, and based on them, the decision can be made regarding the most efficient type (solution) of the thermal protection.

Thus, with the help of dimensional methods, which provide model-prototype correlations through the ML, it becomes possible to replace expensive, cumbersome and difficult-to-repeat investigations on the prototype with sets of experimental measurements performed strictly on the attached model.

The correlation of the measurement data obtained on the model with the prototype behavior prediction is carried out with the help of MLs of different dimensional methods.

It should be noted that the MLs are not identically/uniquely/defined by the different dimensional analysis methods, obviously depending on the assumptions accepted/realized. Obviously, the more accurately the model reflects the behaviors of the prototype, the closer these elements of MLs are to reality.

In the following, the different dimensional methods currently used will be critically analyzed, which offer solutions to approach the propagation of heat that could flow in structural elements.

2. The currently used dimensional methods

2.1 Geometric analogy and similarity theory

In the beginning, the geometric analogy (GA) was used, which assumes a strict geometric similarity between the prototype and model, which significantly restricted its applicability only to simple cases, less applicable to the heat transfer problem [1].

Subsequently, similarity theory (ST) was developed, which, even though it still imposed a geometric similarity, provided solutions for a wider range of problems [2–5]. This was possible, both by the assumption of functional similarity (i.e., similar processes/phenomena occur in the prototype and in the model at similar points and times) and by introducing scale factors, defined for each σ variable involved, in the form of ratios constants, that is

$$S_\sigma = \frac{\sigma_2}{\sigma_1} \, (-) \qquad (1)$$

where always σ_2 refers to the model and σ_1 to prototype.

Also, to the best of the authors' knowledge, problems such as nonlinear or transient heat transfer phenomena being particularly complex can only be solved with the help of real dimensional analysis, analyzed in the following.

2.2 Classical dimensional analysis

For the analysis of complex phenomena, such as that of thermal transfer in structural elements, based on Buckingham's π theorem, hereafter called Classical Dimensional Analysis (CDA), was developed and thoroughly analyzed in the specialized literature [5–19].

In principle, CDA uses a finite number of $\pi_j, j = 1, \ldots, n$ dimensionless variables, which constitute the so-called ML and substitute the set of real variables related to the analyzed phenomenon.

2.3 Shortcoming of the classical dimensional analysis

Even if at first sight CDA would represent the solution for the relevant and efficient analysis of complex phenomena, such as that of heat transfer, we still have to highlight some significant shortcomings, which place this method in the field of the difficult and difficult to achieve by common engineers/specialists.

The first major set of shortcomings results from the very way of obtaining this set of the $\pi_j, j = 1, \ldots, n$ dimensionless variables, which constitute ML.

As is known, this ML can be obtained either:

By direct use of Buckingham's π theorem.

- Through equations with partial differences, which will process the analytical relationship(s) related to the analyzed phenomenon.
- A convenient grouping of the variables from the relations that describe the phenomenon is carried out in order to obtain the set of dimensionless variables.

All these cases require/presuppose a thorough knowledge of higher mathematics, but also of the analyzed phenomenon, which for ordinary engineers/specialists is a difficult criterion to fulfill.

Regarding the first approach, that is, the direct application of Buckingham's theorem, the following shortcomings appear:

- As is well known, in Buckingham's theorem, the set of variables on which the phenomenon depends is defined for the first time. From this crowd, in rather chaotic way (depending on the previous experience of the researcher, or simply involuntary), those variables are chosen (which in MDA we will define, based on a unique-, and rigorous protocol, as independent variables, and which we will place in matrix A, as will be seen in Section 3.1), with the help of which the set of dimensionless variables will be determined, based on the rest of variables in the next step.
- Depending on the exponents of independent variables dimensions (named the primary dimensions, as will be defined in Section 3.1), for the rest of the variables

(which we will define as dependent variables and which will constitute in MDA the matrix B), by solving a system of linear equation, will be determined the corresponding exponents and with their help: the desired/required dimensionless groups; further clustering of these dimensionless entities will eventually produce ML.

- The CDA does not allow the direct highlighting of the independent variables, that is, their conscious placement in the matrix A, precisely defined in MDA. Consequently, CDA cannot ensure the design of flexible models (perfectly adaptable to the intended purpose), that is, models that, based on a protocol of experimental measurements, provide as simple-, safe-, repeatable-, and at a minimal cost the data necessary for the adequate description of the behaviors of the prototype.

- In addition, CDA does not allow the direct introduction into matrix B (so into the set of dependent variables and, thus, into the dimensional set, defined in Section 3.1) of the similarity criteria (or dimensionless standard products) Re, Nu, Gr, etc., which can be included between the elements of ML only either by using their theoretical defining relations or by a rather laborious grouping of the given independent variables. In the latter case, determining the powers/exponents to which these independent variables must intervene requires additional effort on the part of the user.

- The determination of the exponents of the independent variables in order to obtain the ML terms, according to the authors' best knowledge, is done only by the traditional and rather laborious method of solving the system of linear equations, while in MDA, how we will observe in Section 3.1, it results directly, without complicated calculi, based on a simple and straightforward protocol.

- Compared to this, in MDA, if the main goal is pursued in the determination of these similarity criteria (Re, Nu, Gr, etc.), then, also based on the unique protocol, the set of independent variables will be chosen accordingly.

In the case of the other two approaches in obtaining the ML, the following difficulty can be mentioned, namely:

- The methods of forming dimensionless groups based on existing analytical relations, which describe the analyzed phenomenon, limit from the beginning the possible number of these elements (i.e., dimensionless groups), which again contributes to obtaining incomplete MLs.

Also, the identification of this set of $\pi_j, j = 1, \ldots, n$ dimensionless variables are always accompanied by:

- quite chaotic approach, uncertain, difficult and dependent on the previous experience of the respective specialist;

- obtaining full ML only in completely and completely particular cases, precisely because of the few analytical relations we have at our disposal, as well as the way in which Buckingham's theorem is applied;

- the impossibility of ensuring a correlation of the maximum model-prototype flexibility, so necessary for simple, safe, repeatable experimental investigation and at a minimum price of cost per model;
- by the flexibility of the model in relation to the analyzed prototype, we mean finding that solution, that is, that set of independent variables, which ensures the design of a model as simple as possible, useful and with rigorously repeatable experimental investigations;
- the impossibility of eliminating those variables, which have a negligible influence on the respective phenomenon;
- the impossibility of prior selection of those variables that can define an optimal model, that is, the independent variables defined in a unique and safe manner in MDA (see Section 3.2);
- the number of differential equations, respectively, relations in general, that describe the studied phenomenon, is limited, and therefore, obtaining the complete ML is difficult to achieve;
- obtaining the complete ML, in the case of phenomena of great complexity, such as in the case of the heat transfer phenomenon and fire protection, requires in most cases a larger number of dimensionless variables from where the desired ML can be finally obtained.

In the authors' opinion, due to these disadvantages, CDA remains a more theoretical than a practical engineering method (one that is easy to use).

In order to illustrate the application of CDA in thermal problems, a study related to the behavior of RC beams is provided in [20].

In this sense, the author of [20] presents an interesting and original fluid-to-fluid modeling technique in order to model a thermal-hydraulic phenomenon in a working fluid (usually water) by using a special modeling fluid (usually a member of the Freon family). In order to convert the obtained experimental results from Freon (the assigned model) to water (the prototype), the author defined several dimensionless groups, starting from the influencing factors of thermal equilibrium. By applying Buckingham's theorem, were established the main dimensionless groups, the comparison of the predicted and calculated values as well as the obtained uncertainty. The so-obtained and validated ML was applied to other fluids with promising results. The single remark of the authors of this chapter refers to the little bit of a complicated modality/manner in obtaining the demanded ML in comparison with the MDA protocol, as presented in Section 3. This ML can be obtained more easily by predefining the entire set of the influence variables, from where those that have a major influence on the development of the experiments expected on the attached model can be selected. Later, based on the single and simple protocol, by applying the MDA, all the elements of the ML will be easily result.

3. Modern dimensional analysis (MDA)

Compared to these, the dimensional analysis protocol developed by Szirtes in works [21, 22], hereinafter referred to as modern dimensional analysis (MDA), will practically eliminate all the shortcomings of previous methods, which can be easily demonstrated, based on the careful analysis of its stages, (in Section 3.1) and its below-presented facilities (in Section 3.2).

3.1 Calculation steps

First of all, all the variables are chosen, along with their dimensions, which could influence, to a greater or lesser extent, the analyzed phenomenon. It should be mentioned from the beginning that the protocol developed in papers [21, 22] ensures the automatic elimination of variables that do not influence the given phenomenon, which will be detailed later in the analysis of the dimensional set.

From this set of variables, a set of 3 ... 6 variables is chosen, which are directly related to the experimental measurements, which will be performed exclusively on the model attached to the prototype. These variables will be called independent variables, and the dimensions that describe them are primary dimensions.

Bearing in mind that there are nine base units in the International System, only four of which (length, mass, time and temperature) can currently be associated with heat transfer, even if dimension-splitting facilities are used. The dimension splitting is recommended to be applied only with reference to length, because that with mass is particularly dangerous, offering relationships that do not fully satisfy the principles of dimensional analysis [21, 22].

Consequently, in the limiting case (in the best case), we will have a maximum of six primary dimensions $(m_x, m_y, m_z, kg, s, t)$, and correspondingly, a square matrix A with six independent variables and six primary dimensions, so a 6x6 matrix, which is based on the protocol elaborated in the references [21, 22], must necessarily be non-singular, otherwise the basic relation offered by these references cannot be applied, that is, $\det|A| \neq 0$. Moreover, a similar condition, even if not expressly noted, also appears in the CDA, where a linear system of equations must be solved in order to obtain the exponents of the predefined primary dimensions.

The authors, in the framework of theoretical analysis and experiments on the phenomenon of heat transfer, respectively, of fire protection, used the two variants of independent variables, which they presented in Section 4.

Although the authors have not analyzed nonlinear heat transfer problems, one does not see how more variables and dimensions could be involved in the formation/constitution of the A matrix.

In these particularly complex cases, such as those of nonlinear phenomena, the set of independent variables must be chosen very carefully, and the authors are sure that a properly chosen matrix A can solve and must be large enough to describe these nonlinear phenomena as well.

In this sense, it can be mentioned that the author of the contributions [21, 22] solved non-stationary heat transfer problems based on the use of graphs and nomograms with coordinates formed from the elements (and their combinations) of the deduced ML, that is, with dimensionless variables, without exceeding this number of maximum six independent variables.

Practically, for the selection of independent variables, in addition to the fact that they ensure an optimal experimental investigation of the model, only one condition is imposed, namely the matrix A, formed by the exponents of these primary dimensions, must be invertible, so a square matrix with $\det|A| \neq 0$.

If this condition is not fulfilled and we are sure that the chosen independent variables will be the ones that offer an optimal model (so the most flexible from the point of view of the experiences expected exclusively on the attached model), then by changing a line or row we will be able to obtain the desired condition, that the matrix A is non-singular, that is, invertible. The condition that the matrix A, of the independent variables, be invertible is absolutely necessary, as well as in solving a system of linear equations, something that is also done tacitly in CDA to establish the exponents of the variables involved in obtaining the ML.

A remarkable property of these independent variables, which is completely lacking in the rest of the previously presented dimensional methods (i.e., GA, ST and CDA), is that their magnitudes can be chosen a priori, both for the prototype and model.

The following major aspects can be easily observed, namely: it becomes possible not only to choose a model as flexible as possible but also to develop a protocol of experimental investigations on the model as suitable as possible for the intended purpose (production facilities, tests as simple, safe and repeatable as possible and, last but not least, material and human conditions as favorable as possible).

The rest of the variables, for example, n as a number, will constitute the set of dependent variables, and matrix B, formed by the exponents of the primary dimensions of these dependent variables; matrix B can also include several variables with similar dimensions, so without restrictions, as in the case of matrix A.

The size of the dependent variables is chosen a priori only for the prototype, and for the model, they will result exclusively from the ML elements deduced. It should be noted that, among these dependent variables, a limited number refer to the prototype, precisely those whose determination by direct experiment on the prototype would raise particular difficulties and, in fact, for which this dimensional analysis is carried out.

In the case of the latter variables, the quantities related to the model are chosen a priori, and for the prototype, they will result from the deduced ML. This variable selection protocol is specific to MDA and is mandatory in all cases to obtain the expected results. There is NO alternative or other way to adjust this protocol.

Matrices B and A by completing with a specially defined $C = -(A^{-1} \cdot B)^T$ matrix, respectively, with an $D \equiv I_{nxn}$ unitary matrix of order n form the dimensional set defined in the papers [21, 22] and presented in **Table 1**. In the expression of matrix C, the exponent (-1) refers to the inverse matrix, and (T) to the transposed matrix. The n lines, defined from the corresponding lines of matrices D and C, provide a single and simple calculation of the elements of the complete ML, if we carefully selected a priori the complete set of variables, which have a certain influence on the analyzed phenomena.

In this sense, to further illustrate the protocol described in the papers [21, 22], let us assume that ten $(H_1, H_2, \ldots, H_{10})$ variables were identified for the analyzed phenomenon, of which four (H_7, H_8, H_9, H_{10}) main variables with (g_1, g_2, g_3, g_4) dimensions, respectively, $n = 6$ $(H_1, H_2, \ldots, H_6)$ dependent variables.

Matrix B is colored yellow, matrix A green, matrix D red, and C purple.

3.2 Useful notes

- If in the special matrix C, on the column of one of the (H_7, H_8, H_9, H_{10}) main variables, all $(a_1 \ldots \ldots d_6)$ values are zero, then that variable is insignificant for the analyzed phenomenon and must be replaced by another (so it is automatically eliminated from the set of variables initially chosen).

	Dimensions	H_1	H_2	H_3	H_4	H_5	H_6	H_7	H_8	H_9	H_{10}	
B	g_1											A
	g_2											
	g_3											
	g_4											
D	π_1	1	0	0	0	0	0					C
	π_2	0	1	0	0	0	0					
	π_3	0	0	1	0	0	0	a_3	b_3	c_3	d_3	
	π_4	0	0	0	1	0	0					
	π_5	0	0	0	0	1	0					
	π_6	0	0	0	0	0	1					

Table 1.
The dimensional set.

- An element of the complete ML, for example π_3, is obtained based on the relation (2)

$$\pi_3 = (H_3)^1 \cdot (H_7)^{a_3} \cdot (H_8)^{b_3} \cdot (H_9)^{c_3} \cdot (H_{10})^{d_3} = 1, \quad (2)$$

where some of the $(a_3, \dots, d_3)$ exponents can be positive, negative or even.
zero, which is why the final expression of $\pi_3$is a usual algebraic fraction;

- Next, we will perform:

  - Replace the H variables with their S_H scale factors;

- The scale factor of the demanded H_3 variable is expressed, that is,

$$S_{H_3} = \frac{1}{(S_{H_7})^{a_3} \cdot (S_{H_8})^{b_3} \cdot (S_{H_9})^{c_3} \cdot (S_{H_{10}})^{d_3}} = \frac{H_{3,2}}{H_{3,1}}, \quad (3)$$

where $H_{3,2}$ it refers to the model, and $H_{3,1}$ to the prototype, regarding the H_3 dependent variable;

Depending on the type of this H_3 variable (whether it refers to the prototype or the model), one of the magnitudes is known (was chosen) a priori, and the other will result from this element of the ML, how was mentioned/specified above:

- Applying this protocol to the all six $(H_1, H_2, \dots, H_6)$ dependent variables, in the end the full (complete) ML will result, which in the rest of the aforementioned dimensional methods can only be obtained in very, very particular cases.

3.3 The main advantages of MDA

Taking into consideration the previously described protocol, as well as those illustrated in the papers [21, 22], the following main (net) advantages of MDA over the rest of the dimensional methods can be formulated as follows:

- The method does not require thorough knowledge neither in higher mathematics nor in the field of the analyzed phenomenon, requiring only the highlighting of the variables, together with their dimensions, with potential influence on the phenomenon.
- In addition, in the case of the other methods, the selection of independent and dependent variables is not simple and safe, but neither is the exclusion of those with minimal influences; here, at MDA, they are solved directly based on the unique and fast protocol.
- MDA provides, based on an easy, secure (safe)-, and unique protocol to always guarantee full (complete) ML, expert for cases where the obtained dependent variables will not be dimensional expressions and, consequently, cannot be included in the matrix B, based on the MDA protocol described above; in these cases, these variables (more exactly: these dimensionless variables) will be subsequently added to the existing/inferred ML, thus through an empirical adjustment. In practice, as will be shown below, these are only a few special cases.
- In addition, in the case of the other methods, the selection of independent and dependent variables is not simple and safe, but neither is the exclusion of those with minimal influences; here at MDA, they are solved directly based on the unique and fast protocol.
- Regarding the accuracy of MDA compared to the rest of the dimensional methods, if the same elements of ML were obtained, then we have no difference. If, however, a more appropriate (more flexible) ML was obtained with MDA, then the latter provides a better approximation of the results (thus predicting the behavior of the prototype) by means of the ML obtained with MDA.
- From the previous experience of the authors, they could find practically only one limitation of MDA, namely: if even with dimension splitting, we will not obtain dimensional expressions that are adaptable as elements of matrix B, then for them, we will not be able to apply the facilities offered by MDA. Such highlighted cases were:

  Péclet dimensionless number, along the direction $Pe_y = \mathrm{Re}_y \cdot \mathrm{Pr}_y \frac{m_y}{m_x} \cdot \frac{m_x}{m_y} = 1.$

  Stanton dimensionless number, along with directions

$$St_x = \frac{Nu_x}{Pe_x} = Nu_x \cdot \frac{1}{Pe_x} \frac{m_x^2}{m_z \cdot m_y} \cdot \frac{m_z \cdot m_y}{m_x^2} = 1$$

$$St_z = \frac{Nu_z}{Pe_z} = Nu_z \cdot \frac{1}{Pe_z} \frac{m_z^2}{m_x \cdot m_y} \cdot \frac{m_x \cdot m_y}{m_z^2} = 1 \tag{4}$$

when further adjustments such as their subsequent inclusion in ML are recommended based on their theoretical definitional relations.

What the authors mean by model flexibility, explanations were given in Section 2.3, and the net advantages of MDA compared to the rest of the dimensional methods are summarized in Section 3.2.

In the authors' opinion, the MDA practically eliminates the drawbacks of traditional methods. As will be shown in Section 4, the authors performed multiple high-accuracy experimental validations of the deduced model laws based on their own conceived and developed electric test bench. The obtained experimental results, which will be shown, offer a practically total correspondence of the predicted and effective measured values of different variables from the considered prototype.

It allows, due to its special flexibility, the deduction from the general case of some useful particular cases, together with the related simplified models, intended for the analysis of some aspects of great finesse.

It also allows, in order to obtain an ML as simple as possible and with a minimum number of $\pi_j, j = 1, \dots, n$ elements while preserving the complexity of the analyzed phenomenon, the merging of some variables into a new one, respectively, their division, that is, dimensional division, mainly applicable for length and less for mass.

- For the general case, to illustrate the efficiency of combining variables, let us consider the following situation:
  - o At first, the effective dimensions of the cross-section appear in matrix A (e.g., a and b for a rectangular section, or a, b and t for a rectangular-tubular section) when the geometric similarity between the prototype and model is a mandatory condition.
  - o If, however, instead of the actual dimensions, one introduces as an independent variable the A_{tr} area of the cross-section, then the condition of geometric similarity disappears, and it will be possible to choose cross-sections of different shapes for the prototype, respective model.
  - o A much more effective substitution from an engineering point of view is the introduction of the I_{zG} second order of moment of inertia for the cross-section, which more faithfully describes the strength-stress state of the analyzed structural element than its cross-sectional area; so here, too, the condition of geometric similarity will disappear, ensuring the choice of different types of cross-sections for the prototype and model, being imposed only the condition with reference to the adequate (corresponding) $S_{I_{zG}} = \frac{I_{zG,2}}{I_{zG,1}} = const.$, where the index (2) refers to the model and (1) to the prototype.
  - o Going further, if the E Young modulus of elasticity is chosen as an independent variable, and then the prototype and the model can be made (manufactured) of different materials chosen favorably, especially from the point of view of the future experimental investigations that will be performed exclusively on the attached model; in this case, the imposed and unique condition will be obvious $S_E = \frac{E_2}{E_1} = const.$.
  - o If one combines these last two variants, that is, $E \cdot I_{zG}$ the stiffness module is considered as a new independent variable, then neither the shape, the dimensions, nor the material must be identical in the prototype, and in the assigned model, the only condition imposed being obvious $S_{E \cdot I_{zG}} = \frac{E_2 \cdot I_{zG,2}}{E_1 \cdot I_{zG,1}} = const.$; here, then, is what MDA offers special flexibility, which is its exclusive property.
  - o Regarding dimension splitting, mainly applied for length, different (separate) (m_x, m_y, m_z) length scales can be imposed along x, y and z axis

directions, which provides not only reduction (diminishing) of the elements number of ML but also greater flexibility with respect to the attached model; in Chapter 4, this useful aspect will be illustrated.

- If in the desire to simplify the model and follow some particular aspects, then by neglecting/omitting a number of dependent variables from the initial/original Dimensional Set, the rest of the initially inferred ML elements (i.e., corresponding to the remaining dependent variables) remain unchanged.

- The order in which the variables, as well as the involved primary dimensions, are arranged in a matrix (A or B) does not influence the final expressions of the ML elements.

- In the case of the heat transfer phenomenon, especially in the case of fire protection problems, an important (significant) variable that is recommended to be chosen as an independent one is the ς shape factor, defined by the relation (5)

$$\varsigma = \frac{A_{lat}}{V} = \frac{P}{A_{tr}} \left[\frac{1}{m}\right], \tag{5}$$

where A_{lat} represents the lateral area of cross-section;
V represents the volume of the bar;
P represents the perimeter of the cross-section, respectively.
A_{tr} represents the transversal area of the cross-section.

Choosing it as an independent variable offers multiple advantages in the design of particularly efficient and flexible models when the condition of geometric similarity disappears, and it becomes possible to attach models with different cross-sections-compared to the prototype, the imposed condition being

$$S_{\varsigma} = \frac{\varsigma_2}{\varsigma_1} = const. \tag{6}$$

In Chapter 4, based on the authors' own research, the main variables that influence the heat transfer phenomenon, as well as the fire protection issues, will be analyzed. Among these is the Q heat-amount and the corresponding $\dot{Q}$ heat rate, as well as the t, or Δt temperature or the required heating regime.

4. Applications of MDA

To illustrate the great efficiency of MDA, the authors have carried out a series of theoretical research, validated by numerous high-precision experimental investigations, presented in detail in the papers [23–26].

In this regard, to understand the meaning of this somewhat unusual statement, of "high efficiency," the reader is kindly requested to consult the previously mentioned papers, that is, [23–26]. Just in case, these remarkable results, which justified the authors to state this, will be briefly presented below.

First, the conduct of tests in the past, especially on prototypes, was based either on special air-conditioned rooms or on the involvement of large furnaces (usually heated by gas), which raises a number of shortcomings in material, repeatability, but also highly qualified people involved in these investigations.

Considering the fact that the to be deduced ML had to be applied not only to thermally unprotected structural elements but also to those covered with different types of thermo-protective layers (primarily on intumescent paints), the authors performed a series of high-precision tests. The purpose of these tests was to elucidate whether the direction of application of the heat flow on thermally protected structural elements influences or not the size of the thermal conductivity coefficient in the applied thermal protection layer. As is known, usually, this thermal flow is introduced from the outside by means of installations such as special chambers or gas ovens, but the method presents numerous disadvantages, as mentioned before.

If this type of heating could be replaced with one inside the structural element, with the help of specially designed electrical installations, then not only would it increase the efficiency of investigations and reduce costs, but it would also ensure repeatability – but also greater precision in monitoring the amount of heat or the corresponding thermal flow, used for this purpose of heating the respective structural element. After carrying out two types of heating of a specially designed element (i.e., in a classic oven, respectively, by direct heating with an electric current), it was found that the values differ by less than 1%, that is,

$$\lambda = 0.11021 \frac{W}{m \cdot K}, or \frac{W}{m \cdot {}^0C} \text{ (for heating inside)}, \tag{7}$$

$$\lambda = 0.1089 \frac{W}{m \cdot K}, or \frac{W}{m \cdot {}^0C} \text{ (for heating from outside)}, \tag{8}$$

Consequently, even if it might seem curious at first sight, the second method, that is, heating from inside with electric current, presents itself as a better alternative. The authors, based on these results, designed, built and tested an original mobile electric stand (testing bench), three-phase, at 380 V, and with a power of 7 kWh, which can be placed in any ordinary room. This stand presents adequate thermal insulation, that is, during contact in the working area (where the structural element is heated at the natural scale or the models made at the scale 1:2, 1:4 or 1:10) of temperatures of the order of 600^0C, around the stand, at a distance of $0.5m$, the maximum temperature does not exceed 50^0C.

Also, with the help of a specially designed electronic system, temperature stabilization of the order of $500 \dots 600^0C$, based on an original self-adjustment/learning protocol, the initial thermal oscillations compared to the expected nominal values are of $10 \dots 15^0C$, and after 3-4 oscillation cycles, they stabilize at the expected nominal value. Along the structural element (rectangular-tubular bar) subjected to heating/testing, whether it was the prototype or the attached models, they were mounted on all four sides, with the help of miniature screws, thermal-resistances type PT 100-402, with terminals 0.150 m long, with the measurement range of $(-70 \dots +500^0C)$. The average value of the reading of the four thermal resistances at a certain level/height represented the value taken into account for further calculations. The prototype was a segment of a real pillar of an industrial hall, from which a segment of 0.40 m was taken, and according to the geometric analogy models were designed and made on a scale of 1:2, 1:4, and 1:10, respectively. All elements were heated under identical conditions, and for the corresponding times (necessary for the establishing of the homologous times), reaching the same temperatures served as reference points. The number of these thermal resistances was large enough for a proper evaluation of the temperature distribution along these tested elements. The heating of all these elements was carried out at their base, and the achievement of stabilized thermal regimes was taken into account, when at the level of the last thermal resistance (from the upper end) for 120 s, the thermal oscillations did not exceed (0.2 ... 0.3) ^{0}C.

A number of combinations were made, considering the same element as either a prototype or a model. The results of all measurements, which could finally provide either the heat flow or the amount of heat introduced into the tested structural element, were divided into three categories, namely [26]:

- directly measured data, corresponding to the case in which they were linked to a variable of matrix A, that is, they corresponded to an independent variable;
- calculated data, corresponding to the case of variable of matrix B (of the dependent variables), when an element of the ML is applied to determine them, respectively;
- reference data, if they served as data used for comparison with those obtained by calculi with the ML

After careful calculi and checks, practically all values of these two defining variables (Q and $\dot{Q}$) for further experimental investigations overlapped in all cases of prototype-model pairs, regardless of whether they were thermally unprotected or thermally protected. This was also true for the correlated S_Q and $S_{\dot{Q}}$ scales factors.

Apart from these, the rest of the variables involved in the experimental measurements were obviously verified, with a margin of error below 2.5%.

Based on these results, the authors can state that MDA ensures adequate accuracy, accompanied by the reduction of the calculi time and the design time of the ML.

Along with the advantages offered by this original electric stand, experimental investigations can be ensured at a low cost and with reproducibility corresponding to the subsequent statistical evaluation of the measurement data.

Thus, in [25], both the conditions for carrying out experimental investigations as precise and repeatable as possible, as well as the variables that must be included in the analyzed unprotected and thermally protected structural elements (usually with intumescent paints, but not only), were defined. The related experimental measuring stands were also designed and tested under high-precision conditions.

The paper [23] presents the results of theoretical investigations regarding the structural elements of bars with full circular sections, in two versions, that is, not thermally protected and with a thermo-protective layer. MLs were deduced for two cases of great interest, and the facilities offered by them in terms of designing the most flexible models, on which the experimental investigations are to be carried out, were analyzed.

In the papers [24, 26], the authors present, for the case of straight bars with rectangular-tubular (hollow) sections, the results of their theoretical research on the deduction of two useful ML variants, but also the results of experimental measurements to validate these MLs.

In the following, some useful details are provided to those in the field, in the authors' hope that this new approach, through MDA, will gain greater spread and applicability both in the field of theoretical heat transfer studies and especially in the efficiency of the fire protection approach.

A careful analysis of the heat transfer problem in the general case of thermal protected rectangular-tubular cross-sectional beams, depends on the following main variables: heat $Q\,(J)$, time $\tau\,(s)$, heat rate $\dot{Q} = \frac{dQ}{d\tau}\,(W)$, constant-pressure specific heat of air $c_p\left(\frac{J}{{}^0C\cdot kg}\right)$, density (steel, air, paint/insulating material) $\rho\left(\frac{kg}{m^3}\right)$, thermal conductivity (steel, paint coat) along directions *x, y, z* $\lambda_x, \lambda_y, \lambda_z\left(\frac{W}{{}^0C\cdot m}\right)$, specific heat capacity (steel, air) $C\left(\frac{J}{{}^0C}\right)$, velocity normal to the plane where the shear stress is developed $w_0\left(\frac{m}{s}\right)$ with its ∇w_0 gradient, dynamic viscosity of air along directions *x, y, z* $\eta_x, \eta_y, \eta_z\left(\frac{kg}{m\cdot s}\right)$, thermal diffusivity of air along directions *x, y, z* $a_x, a_y, a_z\left(\frac{m^2}{s}\right)$, Prandtl number of air along directions *x, y, z* $\Pr_x, \Pr_y, \Pr_z\,(-)$, kinematic viscosity of air along directions *x, y, z* $\nu_x, \nu_y, \nu_z\left(\frac{m^2}{s}\right)$, convection heat transfer coefficient along directions *x, y, z* $\alpha_{nx}, \alpha_{ny}, \alpha_z\left(\frac{W}{m^2\cdot{}^0C}\right)$, Reynolds number of air along directions *x, y, z* $\mathrm{Re}_x,\ \mathrm{Re}_y,\ \mathrm{Re}_z\,(-)$, volume of beam or paint coat $V\,(m^3)$, thickness of the paint along directions *y, z* $\delta_y, \delta_z\,(m)$, beam's dimensions $L_x, L_y, L_z\,(m)$, lateral area (*x-z*) of the beam $A'_{lat}\,(m^2)$, cross-sectional area of the beam $A_{tr}\,(m^2)$, lateral area (*x-y*) of the beam $A''_{lat}\,(m^2)$, gravitational acceleration $g\left(\frac{m}{s^2}\right)$, shape factor $\varsigma = \frac{A_{lat}}{V} = \frac{P}{A_{tr}}\left(\frac{1}{m}\right)$, cross-sectional perimeter $P\,(m)$, coefficient of volume expansion (steel, air) $\beta\left(\frac{1}{{}^0C}\right)$, temperature variation $\Delta T\,(K)$ or $\Delta t\,({}^0C)$, Grasshoff number along *x* direction $Gr_x\,(-)$, Nusselt number along directions *x, y, z* $Nu_x, Nu_y, Nu_z\,(-)$, Biot number along directions *x, y, z* $Bi_x, Bi_y, Bi_z\,(-)$, Fourier number along directions *x, y, z* $Fo_x, Fo_y, Fo_z\,(-)$, Péclet number along directions *x, y, z* $Pe_x, Pe_y, Pe_z\,(-)$, Stanton number along directions *x, y, z* $St_x, St_y, St_z\,(-)$.

These variables, according to the author's opinion, can be divided into:

- The independent variables, that is, $(L_z, Q, \tau, \Delta t, \varsigma, \lambda_{x\,steel})$, respectively, $(L_z, \dot{Q}, \tau, \Delta t, \varsigma, \lambda_{x\,steel})$ and which will be located in matrix A;

- Dependent variables, which ensure that the experimental measurements are performed under the most favorable conditions (obviously only on the attached model), which are $(L_x, \delta_{y\,steel}, L_y, A'_{lat}, \delta_{z\,steel}, A''_{lat}, A_{tr}, \dot{Q})$, respectively $(L_x, \delta_{y\,steel}, L_y, A'_{lat}, \delta_{z\,steel}, A''_{lat}, A_{tr}, Q)$;

- The dependent variables related to the usual theoretical calculation, that is,

- $\begin{pmatrix} c_{p\,air}, C_{air}, C_{steel}, a_{x\,air}, a_{y\,air}, a_{z\,air}, \rho_{air}, \rho_{steel}, \lambda_{y\,steel}, \lambda_{y\,steel}, \\ \nu_{x\,air}, \nu_{y\,air}, \nu_{z\,air}, \alpha_{nx\,steel}, \alpha_{ny\,steel}, \alpha_{nz\,steel},\ \eta_{x\,air}, \eta_{y\,air}, \eta_{z\,air}, \beta_{air/steel} \end{pmatrix}$;

- The related similarity criteria (the so-called heat transfer correlations), that is $\begin{pmatrix} Crit\,01, Crit\,02, Crit\,03, \Pr_{x\,air}, \Pr_{y\,air}, \Pr_{z\,air}, \\ Gr_{x\,air}, Fo_{x\,air}, Fo_{y\,air}, Fo_{z\,air},\ \mathrm{Re}_{x\,air},\ \mathrm{Re}_{y\,air},\ \mathrm{Re}_{z\,air} \end{pmatrix}$,

where $(Crit01, Crit02, Crit03)$ represent those dimensionless quantities, which had identical/similar expressions and, consequently, the same element of ML, were merged into one entity, that is,

$$Crit\,01 = Nu_z = Pe_z = Bi_z = \frac{m_z^2}{m_x \cdot m_y} \tag{9}$$

$$Crit\,02 = Nu_x = Pe_x = Bi_x = \frac{m_x^2}{m_z \cdot m_y} \tag{10}$$

$$Crit\,03 = Nu_y = St_y = Bi_y = \frac{m_y^2}{m_x \cdot m_z} \tag{11}$$

- The variables, which describe the intumescent paint layer properties or of the other types of thermo-protective layers, that is,

$$\left(\rho_{paint}, \lambda_{x\,paint}, \lambda_{y\,paint}, \lambda_{z\,paint}, \alpha_{nx\,paint}, \alpha_{ny\,paint}, \alpha_{nz\,paint}, \delta_{y\,paint}, \delta_{z\,paint}\right)$$

It should also be noted that the authors focused only on variables that are directly related to the experimental investigations, both for unprotected-, and thermally protected structures.

The rest of variables, that is, those of the theoretical calculi, respectively, related to the similarity criteria, were included in the Dimensional Set only in order to illustrate how easily these variables can be expressed with the help of the set of independent variables chosen, respectively, to be included in the ML thus defined.

It is obvious that, for their establishment on the basis of experimental investigations, the set of independent variables must be reconsidered, that is, the composition of the matrix A; consequently, the elements of the resulting matrix B will also be modified. However, at present, the authors do not have any objective in this sense.

Among these variables, in order to constitute the Dimensional Set, the following were selected as independent variables, namely:

I. $(L_z, Q, \tau, \Delta t, \varsigma, \lambda_{x\,steel})$

II. $\left(L_z, \dot{Q}, \tau, \Delta t, \varsigma, \lambda_{x\,steel}\right)$

In the first case (I.), the heat, and in the second (II.), the heat flow represents the amount of heat introduced into system (in the respective bar/beam).

The two cases (I., II.) were chosen in order to provide researchers with an easier/more convenient option in terms of monitoring the heating of the respective structural element, either by tracking the amount of heat or the heat flow. From the point of view of accuracy, there are no differences between them, if their monitoring was carried out with the same accuracy.

Apart from these, they intervene: L_z one of the lengths, τ time, the Δt thermal regime imposed by the heating rate, the ς shape factor, as well as the $\lambda_{x\ steel}$ thermal conductivity of the steel in the x direction.

Regarding the advantages of these sets of independent variables, taking into account those mentioned in Section 3, the following clarifications can be made:

- Choosing one of the length L_Z ensures, at the specialist's request, that the length scale along the other two x and y directions are identical or different from the one along the z-direction.

- The amount of heat introduced into the bar, both in the prototype and in the model, can be different or identical by choosing the appropriate $S_Q = \frac{Q_2}{Q_1} = const.$ scale factors, respectively, $S_{\dot{Q}} = \frac{\dot{Q}_2}{\dot{Q}_1} = const.$;
- The τ heating time also contribute to ensuring good flexibility of the attached model.
- The thermal regime imposed by Δt also provides significant advantages, because its $S_{\Delta t} = \frac{\Delta t_2}{\Delta t_1} = const.$ scale factor ensures thermal regimes favorable to the model compared to those of the prototype.
- Analyzing these variables, it can be concluded that even if the model is heated to lower temperatures or subjected to milder/reduced thermal regimes, based on the scale factors and ML, a fairly accurate prediction of the behavior of the prototype becomes possible.
- From the combined prototype-model research of the authors [24, 26], it emerged that if the model was heated to at least 400^0 C, then the prediction of the behavior of the prototype, even at temperatures of the order of $800 \ldots 900^0 C$ could be obtained with a smaller margin of error 3.5%.
- This margin of accuracy of 3.5% resulted from the use, in the authors' previously mentioned contributions, of conventional diagrams, where corresponding to different nominal heating temperatures of the tested structural elements (prototype, respectively, models at different scales), these curves, starting from, began to overlap under the error of 3.5%. It refers to the percentage of temperature remaining compared to the nominal one expected at the base of the respective tested element, which was precisely this conventional length. Since the temperature curves will practically coincide at temperatures above, it can be stated that the corresponding thermal regimes but also those related to the state of deformation, will present approximations, if not overlaps of at least this order of magnitudes.
- Since the authors only analyzed steel structural elements of the type of tubular-rectangular bars, they could not verify their validity for other types of materials. It is certain, however, that the ML remains valid for other types of materials since the corresponding variables are among the independent variables used in the two variants I and II.
- Regarding the analysis of other structural configurations, we can state that if the model attached to the prototype will be of the same structural configuration, and the single component elements (of the size of those in the model) subjected to tests will have identical or similar thermal regimes to those of the spatial model attached to the prototype, then we can expect to obtain a similar accuracy.
- Regarding the choice of the $\lambda_{x\,steel}$ thermal conductivity of the steel as an independent variable, with this, the model can be made/manufactured of a material different from the prototype, the imposed condition being the known one $S_{\lambda_{x\,steel}} = \frac{\lambda_{x,2}}{\lambda_{x,1}} = const.$

Obviously, depending on the theoretical or practical purpose pursued, other variables can be chosen as independent variables, but for the authors of this chapter, the applicative part, related to the problem of fire protection, was of interest.

It should also be mentioned that, in the above-mentioned works of the authors, they only looked at the calculation of the elements of ML corresponding to the realization of relevant experimental analysis, both of unprotected-, and thermally protected structural elements; the rest of the elements of ML (those related to some theoretical calculations, respectively, to the establishment of similarity criteria), are just an illustration of how, for a certain set of independent variables, all these relationships useful for a complex analysis can be determined as simply and uniformly as possible.

As was mentioned, if we propose such a theoretical investigation, then, taking into account the variables whose influence is decisive for these cases, they will be included in the set of independent variables, obviously, with the only mandatory condition, that the matrix A thus constituted is invertible, so a square matrix with $\det|A| \neq 0$.

To illustrate the application of these two sets of independent variables, some useful results for those in the field are presented below.

For case I, the matrix A, formed by the exponents of the $(L_z, Q, \tau, \Delta t, \varsigma, \lambda_{x\,steel})$ variables, leads to the form shown in **Table 2**.

In this table (m_x, m_y, m_z) it refers to dimension splitting of the length mentioned above, actually to obtain a best-fit ML. To perform a relevant experimental analysis, the authors suggest that matrix B consists of the variables mentioned/chosen in **Table 3**.

By performing the adequate protocol, based on relations (2) and (3), finally one can obtain the corresponding ML, that is:

$$\pi_1 = \dot{Q} \cdot Q^{-1} \cdot L_z^0 \cdot \Delta t^0 \cdot \tau^1 \cdot \lambda_{x\,ot}^0 \cdot \varsigma^0 = \frac{\dot{Q} \cdot \tau}{Q} = 1 \Rightarrow \frac{S_{\dot{Q}} \cdot S_\tau}{S_Q} = 1 \Rightarrow S_{\dot{Q}} = \frac{S_Q}{S_\tau} \tag{12}$$

$$\pi_2 : S_{A_{tr}} = \frac{S_{L_z}}{S_\varsigma} \tag{13}$$

$$\pi_3 : S_{A'_{lat}} = \frac{S_Q \cdot S_{L_z}}{S_{\Delta t} \cdot S_\tau \cdot S_{\lambda_{x\ steel}}} \tag{14}$$

$$\pi_4 : S_{A''_{lat}} = \frac{S_Q}{S_{\Delta t} \cdot S_\tau \cdot S_{\lambda_{x\ steel}} \cdot S_\varsigma} \tag{15}$$

$$\pi_5 : S_{L_x} = \frac{S_Q}{S_{\Delta t} \cdot S_\tau \cdot S_{\lambda_{x\ steel}}} \tag{16}$$

Dimensions	L_z	Q	τ	Δt	ς	$\lambda_{x\,steel}$
m_x	0	2	0	0	0	1
m_y	0	0	0	0	−1	0
m_z	1	0	0	0	0	0
kg	0	1	0	0	0	1
s	0	−2	1	0	0	−3
°C	0	0	0	1	0	−1

Table 2.
Matrix A for the case I.

Dimensions	L_x	$\delta_{y\,steel}$	L_y	A'_{lat}	$\delta_{z\,steel}$	A''_{lat}	A_{tr}	$\dot{Q}$
m_x	1	0	0	1	0	1	0	2
m_y	0	1	1	0	0	1	1	0
m_z	0	0	0	1	1	0	1	0
kg	0	0	0	0	0	0	0	1
s	0	0	0	0	0	0	0	−3
$°C$	0	0	0	0	0	0	0	0

Table 3.
Matrix B with the experimental part.

$$\pi_6 : S_{L_y} = \frac{1}{S_\varsigma} \tag{17}$$

$$\pi_7 : S_{\delta_{y\ steel}} = \frac{1}{S_\varsigma} \tag{18}$$

$$\pi_8 : S_{\delta_{z\ steel}} = S_{L_z} \tag{19}$$

From a careful analysis of relations (12)–(19), it can be seen that, in the case of choosing similar materials and testing conditions for the prototype and the model, these expressions are significantly simplified, as will be detailed below. The mentioned simplifications will be valid for all involved scale factors without other changes.

Also, if one wishes to accept the same scale of all lengths, as defined by S_{L_z}, then some independent variables in matrix B, listed in **Table 3**, can simply be ignored. In this sense, for the case of the rectangular-tubular cross-section, the square-tubular one represents a special case without the need to derive another ML.

Similarly, for the second case (II.) $\left(L_z, \dot{Q}, \tau, \Delta t, \varsigma, \lambda_{x\,steel}\right)$, they are defined one at a time:

- The independent variables (see **Table 4**)

- The relevant variables for the experimental investigations (see **Table 5**)

- The corresponding elements of ML, offered by relations (20)–(27)

$$\pi_1 = Q \cdot \dot{Q}^{-1} \cdot L_z^0 \cdot \Delta t^0 \cdot \tau^{-1} \cdot \lambda_{x\ steel}^0 \cdot \varsigma^0 = \frac{Q}{\dot{Q} \cdot \tau} = 1 \Rightarrow S_Q = S_{\dot{Q}} \cdot S_\tau \tag{20}$$

$$\pi_2 : S_{A_{tr}} = \frac{S_{L_z}}{S_\varsigma} \tag{21}$$

$$\pi_3 : S_{A'_{lat}} = \frac{S_{\dot{Q}} \cdot S_{L_z}}{S_{\Delta t} \cdot S_{\lambda_{x\ steel}}} \tag{22}$$

$$\pi_4 : S_{A''_{lat}} = \frac{S_{\dot{Q}}}{S_{\Delta t} \cdot S_{\lambda_{x\ steel}} \cdot S_\varsigma} \tag{23}$$

Dimensions	L_z	$\dot{Q}$	τ	Δt	ς	$\lambda_{x\,steel}$
m_x	0	2	0	0	0	1
m_y	0	0	0	0	−1	0
m_z	1	0	0	0	0	0
kg	0	1	0	0	0	1
s	0	−3	1	0	0	−3
°C	0	0	0	1	0	−1

Table 4.
Matrix A for the case II.

Dimensions	L_x	$\delta_{y\,steel}$	L_y	A'_{lat}	$\delta_{z\,steel}$	A''_{lat}	A_{tr}	Q
m_x	1	0	0	1	0	1	0	2
m_y	0	1	1	0	0	1	1	0
m_z	0	0	0	1	1	0	1	0
kg	0	0	0	0	0	0	0	1
s	0	0	0	0	0	0	0	−2
°C	0	0	0	0	0	0	0	0

Table 5.
Matrix B with the experimental part.

$$\pi_5 : S_{L_x} = \frac{S_{\dot{Q}}}{S_{\Delta t} \cdot S_{\lambda_{x\ steel}}} \tag{24}$$

$$\pi_6 : S_{L_y} = \frac{1}{S_\varsigma} \tag{25}$$

$$\pi_7 : S_{\delta_{y\ steel}} = \frac{1}{S_\varsigma} \tag{26}$$

$$\pi_8 : S_{\delta_{z\ steel}} = S_{L_z} \tag{27}$$

Regarding the applicability of the elements of ML to other materials, the answer is affirmative, since the independent variables include the coefficient of thermal conductivity, so that these relations can be generalized with reference to the material of the structural element.

Regarding the structural configuration, if the experimental investigations to which the model will be subjected are manufactured with a structure having the same configuration as the prototype, again, the elements of ML can be applied.

In addition, one can mention the following:

- If one has identical materials for prototype and model, then relation (28)

$$S_{\alpha_{nx\,steel}} = S_{\alpha_{nr\,steel}} = S_{\lambda_{x\,steel}} = S_{\lambda_{r\,steel}} = S_{\rho_{steel}} = S_{C_{steel}} = S_{\beta_{steel}} = 1 \tag{28}$$

brings further simplifications;

- Identical conditions of experimental investigations lead to the simplifications provided by the relation (29)

$$S_{c_{p\,air}} = S_{C_{air}} = S_{a_{x\,air}} = S_{a_{r\,air}} = S_{\rho_{air}} = S_{\eta_{x\,air}} = S_{\eta_{r\,air}} = S_{\nu_{x\,air}} = S_{\nu_{r\,air}} = S_{\beta_{air}} = 1 \quad (29)$$

- If the same type of thermal protection layer was used under identical conditions, then relation (30) should be considered

$$S_{\rho_{paint}} = S_{\lambda_{x\,paint}} = S_{\lambda_{r\,paint}} = S_{\alpha_{nr\,paint}} = 1 \quad (30)$$

- Finally, accepting the same scale of all lengths, relation (31) will lead to other major simplifications of ML, such as among others, that the dependent variables related to the initially defined $(A'_{lat}, A''_{lat}, A_{tr})$ areas and their scale factors can be completely ignored/omitted from the analysis.

$$S_{L_x} = S_{L_y} = S_{L_z} = ct. \quad (31)$$

Obviously, in the case of materials with different properties along certain directions or in anisotropic ones, this simplification must be reconsidered and will be applied with caution only based on preliminary experimental investigations regarding the influence of these inhomogeneities in behavior. In the case of non-uniformly heated structures, the same non-uniformities should be kept as much as possible in the model under experimental investigation with those of the analyzed prototype, even though both time, temperature/thermal regime, and heat (or input heat flux) are independent variables, so that they can be chosen a priori for both the prototype and the model.

Of course, based on the experience gained by those in the field, they will be able to improve this dimensional analysis based on the principles of MDA established in works [21, 22].

5. Additional facilities of MDA

The question can rightly be asked whether these results of MDA, that is, inferred ML, obtained for simple structural elements can be extended to complex structures such as buildings, but not only.

The inferred ML can be extended to complex structures if:

- consist of those types of structural elements (identical configurations) that have been tested for the validation of the ML;
- their materials are the same as those of the model subjected to validation tests, at least in type, that is, homogeneous, isotropic or orthotropic;
- the test conditions are identical to those of the model.

The answer is favorable, especially due to the fact that MDA also operate, based on the assumption of functional similarity, with phenomena that occur in homologous times in homologous points of the prototype and the attached model.

In this sense, given the fact that, among the independent variables of the approach proposed by the authors, in Section 4, there are both the τ time, t_0 temperature (or the temperature difference related to the Δt_0 thermal regime imposed in the two structures), and the L_z geometric dimension, based on some conveniently chosen scale factors, the homologous points, the homologous times, but also the temperatures (as studied phenomenon) that will be reached in the homologous points of the two structures (prototype, respectively, model) can be easily identified.

Regarding the convenient choice of scale factors, the following major elements can be noted from the authors' experience:

- for length, S_L must be chosen so that the model obtained is:
    - large enough (to allow the placement of an adequate number of thermal resistances or thermocouples);
    - found in the range of standardized structural elements (e.g., in the case of profiles I, U, Z etc.);
- suitable for tests with acceptable power consumption:
    - for example, if the heating of the structural element with electric current is applied, a too bulky model requires a bulky installation with high energy performance, which would require a significant material effort; in addition, the thermal inertia would be too high, so stabilizing the temperature of the heated model would raise problems;
    - if the model were miniature, then the temperature monitoring conditions would be more difficult, but also an excessive influence of environmental factors on the temperature and its stability;
- for the temperature, S_T, this temperature of the model should be as close as possible to that of the prototype, but at least 400 ... 450^0C, so that based on some curves with conventional scales (such as those used by the authors, $T_\Psi - L_\Psi$) the best forecast of the temperature of the prototype is obtained; obviously, a higher thermal regime of the tested model also imposes additional fire protection conditions for the installation and the premises where the experiment will be carried out;
- for the material, S_λ, the same type of material is preferred, if the dimensions for the model are found among the standardized structural elements; also, even if another material is chosen for the model, it should be of the same type (isotropic, orthotropic, etc.), with fairly predictable values for λ with increasing temperature, otherwise additional adjustments of the behaviors of the model with the temperature variation will be required;
- for the S_ς shape factor, models whose heating takes place on an identical number of sides to those of the prototype are preferred; the values will result according to the accepted dimensions of the model;
- for S_Q or $S_{\dot{Q}}$, these Q or $\dot{Q}$ values, related to the model, must not lead to structural changes in the material, nor to changes in the load-bearing capacity of the model (if it will be subjected to combined thermal and mechanical stresses);
- for time, S_τ, these exposure times of the model depend on the previous factors, and based on some preliminary tests, the optimal heating times can be found;

- in any case, to be able to identify the homologous times of the model with the prototype, stabilization times are recommended (to reach some imposed/predefined nominal temperatures).

The specialist only needs to add a t_0^* temperature to the dependent variables, for which, the corresponding element in the ML, can be subsequently applied to the three directions x, y and z, that is, to define the behavior in this sense, that is, the temperatures $\left(t_x^*,\ t_y^*,\ t_z^*\right)$ of some homologous points of the structural elements in the 3D space.

Thus, the first time the appropriate model is chosen for the prototype, respectively, the time-, temperature-, and length scale factors are defined.

If it is a spatial structure, such as the resistance structure of a building or an industrial hall, taking into account temperature regime define by $S_{t_0} = \frac{t_{0,2}}{t_{0,1}} = const.$ the scale factor, then high-precision measurements will be made/performed on the corresponding model, still being a spatial structure, with similar shape/form to the prototype (having the same number of pillars and crossbars, disposed similarly).

At the beginning of Section 4, the thermal regime monitoring protocol was presented with the help of PT 100-402 thermal resistances, which the authors previously calibrated individually with the help of precision thermocouples. With the help of these calibration curves the measurement results were corrected accordingly.

Without caring out the actual deduction of the corresponding element of the ML for this t_0^* temperature, it will finally result, depending, for example, on the imposed Δt_0 thermal regime), the relation (32):

$$\begin{aligned} &\pi_k = (t^*)^1 \cdot (\Delta t_0)^{-1} = 1 \Rightarrow t^* = \Delta t_0 \Rightarrow \\ &\Rightarrow S_{t*} = S_{\Delta t_0};\ \text{with } t^* = t_x^*, t_y^*, t_z^*. \end{aligned} \tag{32}$$

This relation (32) is applicable in all homologous points of the two structures, obviously in homologous times, defined by the $S_\tau = \frac{\tau_2}{\tau_1} = const.$ time scale factor.

It should be noted that relation (29) is an element of the extended ML and is not proved/deduced in this section. It does not represent a new analytical relation, but, according to the MDA protocol, is only a correlation between the independent (here) and dependent variables involved (here), which was deduced according to relations (2) and (3) of Section 3.1., so that finally it provides a model-prototype correlation through ML.

Having the values from the results of the measurements carried out on the attached model, that is $t_{x,2}^*, t_{y,2}^*, t_{z,2}^*$, the values of the homologous points of the prototype, that is, the probable $t_{x,1}^*, t_{y,1}^*, t_{z,1}^*$ temperatures of the corresponding points on the real structure, will be simply obtained with the help of the relation (33)

$$\frac{t_2^*}{t_1^*} = S_{\Delta t_0} \Rightarrow t_1^* = \frac{t_2^*}{S_{\Delta t 0}}, \text{for} x, y, z. \tag{33}$$

The relation (33) was used by the authors in steady-state conditions, but if precise identification of the homologous times is made for the two structures (prototype and model), then it also becomes applicable in the case of transient thermal phenomena.

Regarding the establishment of homologous times, the authors used in their previous investigations the times corresponding to reaching the same stabilized thermal regimes, based on rigorous criteria previously presented in Section 4.

In order to facilitate the identification of the homologous points on the two structures (prototype, respectively, model), it is proposed to introduce the $L_{\%}$ conventional length. By definition, it will represents the percentage of the L_x, L_y or L_z total length of a certain elevation on the respective structural element (prototype or model). Thus, according to the authors, it becomes easier to identify these homologous points.

In the analyzed case, being two symmetrical structures, of the same configuration, the identification of the homologous points only requires the most accurate establishment of the corresponding position relative to a certain on the model compared to the prototype. If it were an asymmetric structure, then, as previously mentioned, both the prototype and the model should have the same configuration (same type of geometric asymmetry); otherwise, in the author's opinion, such identification of homologous points becomes impossible.

From the previous experience of the authors, as mentioned in Section 4, if the model was tested at thermal regimes of at least 400°C, then with a maximum 3.5% error, the behavior of the prototype can be predicted at 800 ... 900°C, which already corresponds to real fire cases, or a little close by them.

6. Conclusions

- In order to study complex phenomena as simply and efficiently as possible, the specialists in the field, together with mathematicians, have developed different approaches. Among them, the various dimensional methods have gained special importance.

- Taking into account the fact that both thermal transfer in structural elements and the issue of fire protection of buildings made from them are phenomena of great complexity, in this chapter, the authors have made a synthesis and a critical analysis of the dimensional methods currently used in these fields, highlighting their advantages and disadvantages (limits/shortcomings).

- Geometric analogy is a very simple and easy approach, but it has two major limitations: the mandatory condition of geometric similarity and the small number of dimensionless relations/groups, which can be obtained by normalization from the available variables.

- The Similarity Theory allows solving somewhat more complicated problems by assuming functional similarity but has geometric similarity as a mandatory condition.

- The Classical Dimensional Analysis in the first approach is the solution for the analysis of complex phenomena, which operates with numerous dimensionless groups. However, it has a number of major shortcomings, such as: the three cumbersome ways of establishing these dimensionless groups; the impossibility of directly selecting, based on an unique protocol, the independent variables that will define both the attached model and its test conditions, and thus: the design of a flexible model; the impossibility of directly introducing the criteria relations (similarity criteria) into the Dimensional Set and thus obtaining a complete set of independent variables, to be fully expressed as a function of the chosen independent variables; the rather cumbersome determination of the exponents of the independent variables to deduce the ML; the reduced number of analytical relations that describe the analyzed phenomenon, from which the elements of the ML can be deduced, that is, of the set of dimensionless variables; obtaining the complete ML is achieved only in particular cases; in conclusion, in the opinion of the authors, CDA remains a much more theoretical than simple engineering method within the reach of anyone.

- Compared to these, the MDA brief presented, but hopefully under conditions of immediate applicability, provides a safe-, simple-, and accessible way for all researchers and engineers in the field.

- Based on what was analyzed and illustrated in Sections 3 and 4, MDA practically eliminates all the shortcomings of the other dimensional methods and represents, in the opinion of the authors, an accessible, engineering method.

- Thus, after presenting the principles governing the MDA for the case of the thermal transfer phenomenon in steel structural elements, the method of obtaining two MLs is illustrated with reference to the steel beams with a rectangular hole cross-section in unprotected-, and thermally protected versions, whose applicability and the validation by high-accuracy experimental measurements was verified by the authors.

- After deduction of these two MLs, the authors carried out detailed experimental investigations, detailed in Section 4. The purpose of these investigations was a rigorous validation of the two MLs, when on four elements, made at the scale of 1:1, 1:2, 1:4 and 1:10, measurements of the basic quantities governing heat transfer were made. In this sense, the four structural elements tested had either the role of a prototype or a model, with similar heating tests being performed, with the monitoring of the involved parameters. Regarding the three categories of data (directly measured, calculated, respectively, reference data), at $\dot{Q}$ and Q, respectively, their $S_{\dot{Q}}$ and S_Q scale factors, a perfect overlap of them was obtained, while for the rest of the variables involved in the experimental measurements, an error margin below 2.5% was obtained; in [26] the authors detailed these measurements. In the case of extending these MLs to spatial structures (model and prototype to a four-compartments industrial hall), by means of conventional $T_\Psi - L_\Psi$ diagrams, where heating of the attached model was done/performed to at least 400^0C, the margin of error was 3.5%.

- In this sense, based on the previous experience of the authors, the simplest ways to obtain flexible models are presented, which also ensure rigorous-, repeatable-, simple-, and safe experimental measurements on the attached model.

- Useful details are provided on how, with the help of variable merging (variable combining) and variable splitting, respectively, those in the field can obtain MLs as simply and safely as possible.

- It should also be emphasized that the detailed step-by-step analysis of independent as well as dependent variables will be of great help to those who are just beginning to apply this simple and safe method in engineering practice, that is, MDA.

- Although in this chapter, the ML elements related to the theoretical aspects, the criterion elements (parameters), respectively, those related to the thermal protection layers are not analyzed, they can be found in the previous works of the authors [23–26].

- An important aspect, according to the authors, is also related to how these MLs can be applied to 3D structures, based on the hypothesis of functional similarity, with phenomena occurring at homologous times in homologous points of the prototype and the attached model; this last aspect comes to underline the net advantages of MDA over the rest of the dimensional methods.

- To the best of the authors' knowledge, these detailed aspects of the heat transfer problem in structural members such as rectangular-tubular cross-sectional steel bars, based on MDA principle, have not been presented by other authors.

The variables that can have an influence on the heat transfer in the beam with rectangular or rectangular hole section (**Table 6**).

	Variable	
	Symbol/formula	**Dimension**
Heat* (J)	Q	$J = N \cdot m = \frac{kg \cdot m_x}{s^2} \cdot m_x = \frac{kg \cdot m_x^2}{s^2}$
Heat rate (W)	$\dot{Q} = \frac{dQ}{d\tau}$	$W = \frac{J}{s} = \frac{kg . m_x^2}{s^3}$
Time (s)	τ	s
Density $\frac{kg}{m^3}$ (steel, air, paint/ insulating material)	ρ	$\frac{kg}{m^3} = \frac{kg}{m_x \cdot m_y \cdot m_z}$
Constant-pressure specific heat of air $\left(\frac{J}{K \cdot kg}\right)$	$c_p = \frac{1}{m} \cdot \frac{dQ}{dt}$	$\frac{1}{kg} \cdot \frac{J}{K} = \frac{1}{kg} \cdot \frac{kg \cdot m_x^2}{s^2 \cdot K} = \frac{m_x^2}{s^2 \cdot K}$;
Specific heat capacity (steel, air) $\left(\frac{J}{K}\right)$	$C = \frac{dQ}{dT}$	$\frac{J}{K} = \frac{kg \cdot m_x^2}{s^2 \cdot K}$
Thermal conductivity (steel, paint coat), along directions x,y,z $\left(\frac{W}{K \cdot m}\right)$	λ_x	$\frac{W}{m_x \cdot K} = \frac{kg \cdot m_x}{s^3 \cdot K}$
	λ_y	$\frac{W}{m_y \cdot K} = \frac{kg \cdot m_x^2}{m_y \cdot s^3 \cdot K}$
	λ_z	$\frac{W}{m_z \cdot K} = \frac{kg \cdot m_x^2}{m_z \cdot s^3 \cdot K}$
Thermal diffusivity of air, along directions x,y,z (m^2/s)	$a_x = \frac{\lambda_x}{\rho \cdot c_p} = \frac{1}{\rho} \cdot \frac{1}{c_p} \cdot \lambda_x$	$\frac{m_x \cdot m_y \cdot m_z}{kg} \cdot \frac{s^2 \cdot K}{m_x^2} \cdot \frac{kg \cdot m_x}{s^3 \cdot K} = \frac{m_y \cdot m_z}{s}$
	$a_y = \frac{\lambda_y}{\rho \cdot c_p} = \frac{1}{\rho} \cdot \frac{1}{c_p} \cdot \lambda_y$	$\frac{m_x \cdot m_y \cdot m_z}{kg} \cdot \frac{s^2 \cdot K}{m_x^2} \cdot \frac{kg \cdot m_x^2}{m_y \cdot s^3 \cdot K} = \frac{m_x \cdot m_z}{s}$
	$a_z = \frac{\lambda_z}{\rho \cdot c_p} = \frac{1}{\rho} \cdot \frac{1}{c_p} \cdot \lambda_z$	$\frac{m_x \cdot m_y \cdot m_z}{kg} \cdot \frac{s^2 \cdot K}{m_x^2} \cdot \frac{kg \cdot m_x^2}{m_z \cdot s^3 \cdot K} = \frac{m_x \cdot m_y}{s}$
Velocity **(m/s)	w_0	$\frac{m}{s}$
Dynamic viscosity of air, along directions x, y,z $\left(\frac{kg}{m \cdot s}\right)$	$\eta_x = \frac{\tau_{0,x}}{\nabla w_0} = \tau_{0,x} \cdot \frac{1}{\nabla w_0} = \frac{F_{0,x}}{A_{xy}} \cdot \frac{1}{\nabla w_0}$	$\frac{kg}{s^2 \cdot m_y} \cdot \frac{1}{1/s} = \frac{kg}{s \cdot m_y}$
	$\eta_y = \frac{\tau_{0,y}}{\nabla w_0} = \tau_{0,y} \cdot \frac{1}{\nabla w_0} = \frac{F_{0,y}}{A_{zy}} \cdot \frac{1}{\nabla w_0}$	$\frac{kg}{s^2 \cdot m_z} \cdot \frac{1}{1/s} = \frac{kg}{s \cdot m_z}$
	$\eta_z = \frac{\tau_{0,z}}{\nabla w_0} = \tau_{0,z} \cdot \frac{1}{\nabla w_0} = \frac{F_{0,z}}{A_{yz}} \cdot \frac{1}{\nabla w_0}$	$\frac{kg}{s^2 \cdot m_y} \cdot \frac{1}{1/s} = \frac{kg}{s \cdot m_y}$

	Variable	
	Symbol/formula	**Dimension**
Kinematic viscosity of air, along directions x,y,z (m^2/s)	$\nu_x = \frac{\eta_x}{\rho} = \frac{1}{\rho} \cdot \eta_x$	$\frac{m_x \cdot m_y \cdot m_z}{kg} \cdot \frac{kg}{s \cdot m_y} = \frac{m_x \cdot m_z}{s}$
	$\nu_y = \frac{\eta_y}{\rho} = \frac{1}{\rho} \cdot \eta_y$	$\frac{m_x \cdot m_y \cdot m_z}{kg} \cdot \frac{kg}{s \cdot m_z} = \frac{m_x \cdot m_y}{s}$
	$\nu_z = \frac{\eta_z}{\rho} = \frac{1}{\rho} \cdot \eta_z$	$\frac{m_x \cdot m_y \cdot m_z}{kg} \cdot \frac{kg}{s \cdot m_y} = \frac{m_x \cdot m_z}{s}$
Prandtl dimensionless number of air, along directions	$\mathrm{Pr}_x = \frac{\nu_x}{a_x} = \nu_x \cdot \frac{1}{a_x}$	$\frac{m_x \cdot m_z}{s} \cdot \frac{s}{m_y \cdot m_z} = \frac{m_x}{m_y}$
	$\mathrm{Pr}_y = \frac{\nu_y}{a_y} = \nu_y \cdot \frac{1}{a_y}$	$\frac{m_x \cdot m_y}{s} \cdot \frac{s}{m_x \cdot m_z} = \frac{m_y}{m_z}$
	$\mathrm{Pr}_z = \frac{\nu_z}{a_z} = \nu_z \cdot \frac{1}{a_z}$	$\frac{m_x \cdot m_z}{s} \cdot \frac{s}{m_y \cdot m_x} = \frac{m_z}{m_y}$
Reynolds dimensionless number, along directions	$\mathrm{Re}_x = \frac{w_{0,x} \cdot l_x}{\nu_x} = w_{0,x} \cdot l_x \cdot \frac{1}{\nu_x}$	$\frac{m_x}{s} \cdot m_x \cdot \frac{s}{m_x \cdot m_z} = \frac{m_x}{m_z}$
	$\mathrm{Re}_y = \frac{w_{0,y} \cdot l_y}{\nu_y} = w_{0,y} \cdot l_y \cdot \frac{1}{\nu_y}$	$\frac{m_y}{s} \cdot m_y \cdot \frac{s}{m_x \cdot m_y} = \frac{m_y}{m_x}$
	$\mathrm{Re}_z = \frac{w_{0,z} \cdot l_z}{\nu_z} = w_{0,z} \cdot l_z \cdot \frac{1}{\nu_z}$	$\frac{m_z}{s} \cdot m_z \cdot \frac{s}{m_x \cdot m_z} = \frac{m_z}{m_x}$
Convection heat transfer coefficient, along directions x,y,z *** $\left(\frac{W}{K \cdot m^2}\right)$	α_{ny}	$\frac{W}{m^2 \cdot K} = \frac{kg \cdot m_x^2}{s^3} \cdot \frac{1}{m_x \cdot m_z \cdot K} = \frac{kg \cdot m_x}{s^3 \cdot m_z \cdot K}$
	α_{nz}	$\frac{W}{m^2 \cdot K} = \frac{kg \cdot m_x^2}{s^3} \cdot \frac{1}{m_x \cdot m_y \cdot K} = \frac{kg \cdot m_x}{m_y \cdot s^3 \cdot K}$
	α_{nx}	$\frac{W}{m^2 \cdot K} = \frac{kg \cdot m_x^2}{s^3} \cdot \frac{1}{m_z \cdot m_y \cdot K} = \frac{kg \cdot m_x^2}{m_z \cdot m_y \cdot s^3 \cdot K}$
Thickness of the paint coat, along directions x,y,z (m)	$d_y = \delta_y$	m_y
	$d_z = \delta_z$	m_z
Volume of bar or paint coat (m^3)	V	$m^3 = m_x \cdot m_y \cdot m_z$
Area of the bar cross-section (m^2)	A_{tr}	$m_y \cdot m_z$
Lateral area *(x-z)* (m^2)	A'_{lat}	$m_x \cdot m_z$
Lateral area *(x-y)* (m^2)	A''_{lat}	$m_x \cdot m_y$
Bar dimensions (m)	L_x	m_x
	L_y	m_y
	L_z	m_z
Shape factor (m^{-1})	$\varsigma = \frac{A_{lat}}{V} = \frac{P}{A_{tr}}$; P is the cross-section perimeter	$\frac{1}{m_y}$
Gravitational acceleration (m/s^2)	g	$\frac{m}{s^2} = \frac{m_x}{s^2}$
Temperature variation: (K)	ΔT	K
Coefficient of volume expansion (steel, air) $(K)^{-1}$	β	$\frac{1}{K}$
Nusselt dimensionless number, along directions	$Nu_x = \frac{\alpha_x \cdot l_x}{\lambda_{f,x}} = \alpha_x \cdot l_x \cdot \frac{1}{\lambda_{f,x}}$	$\frac{kg \cdot m_x^2}{s^3 \cdot m_y \cdot m_z \cdot K} \cdot m_x \cdot \frac{s^3 \cdot K}{kg \cdot m_x} = \frac{m_x^2}{m_z \cdot m_y}$
	$Nu_y = \frac{\alpha_y \cdot l_y}{\lambda_{f,y}} = \alpha_y \cdot l_y \cdot \frac{1}{\lambda_{f,y}}$	$\frac{kg \cdot m_x}{s^3 \cdot m_z \cdot K} \cdot m_y \cdot \frac{m_y \cdot s^3 \cdot K}{kg \cdot m_x^2} = \frac{m_y^2}{m_z \cdot m_x}$
	$Nu_z = \frac{\alpha_z \cdot l_z}{\lambda_{f,z}} = \alpha_z \cdot l_z \cdot \frac{1}{\lambda_{f,z}}$	$\frac{kg \cdot m_x}{s^3 \cdot m_y \cdot K} \cdot m_z \cdot \frac{m_z \cdot s^3 \cdot K}{kg \cdot m_x^2} = \frac{m_z^2}{m_x \cdot m_y}$
Grashof dimensionless number, along direction	$Gr_x = \frac{g \cdot \beta \cdot \Delta t \cdot l_\zeta^3}{\nu_\zeta^2} = g \cdot \beta \cdot \Delta t \cdot l_x^3 \cdot \frac{1}{\nu_x^2}$	$\frac{m_x}{s^2} \cdot \frac{1}{K} \cdot K \cdot m_x^3 \cdot \frac{s^2}{m_x^2 \cdot m_z^2} = \frac{m_x^2}{m_z^2}$

	Variable	
	Symbol/formula	**Dimension**
Péclet dimensionless number, along directions	$Pe_x = \mathrm{Re}_x \cdot \mathrm{Pr}_x$	$\frac{m_x}{m_z} \cdot \frac{m_x}{m_y} = \frac{m_x^2}{m_z \cdot m_y}$
	$Pe_y = \mathrm{Re}_y \cdot \mathrm{Pr}_y$	$\frac{m_y}{m_x} \cdot \frac{m_x}{m_y} = 1^{****}$
	$Pe_z = \mathrm{Re}_z \cdot \mathrm{Pr}_z$	$\frac{m_z}{m_x} \cdot \frac{m_z}{m_y} = \frac{m_z^2}{m_x \cdot m_y}$
Biot dimensionless number, along directions	$Bi_x = \frac{\alpha_x \cdot l_x}{\lambda_{s,x}} = \alpha_x \cdot l_x \cdot \frac{1}{\lambda_{s,x}}$	$\frac{kg \cdot m_x^2}{s^3 \cdot m_y \cdot m_z \cdot K} \cdot m_x \cdot \frac{s^3 \cdot K}{kg \cdot m_x} = \frac{m_x^2}{m_z \cdot m_y}$
	$Bi_y = \frac{\alpha_y \cdot l_y}{\lambda_{s,y}} = \alpha_y \cdot l_y \cdot \frac{1}{\lambda_{s,y}}$	$\frac{kg \cdot m_x}{s^3 \cdot m_z \cdot K} \cdot m_y \cdot \frac{s^3 \cdot m_y \cdot K}{kg \cdot m_x^2} = \frac{m_y^2}{m_z \cdot m_x}$
	$Bi_z = \frac{\alpha_z \cdot l_z}{\lambda_{s,z}} = \alpha_z \cdot l_z \cdot \frac{1}{\lambda_{s,z}}$	$\frac{kg \cdot m_x}{s^3 \cdot m_y \cdot K} \cdot m_z \cdot \frac{s^3 \cdot m_z \cdot K}{kg \cdot m_x^2} = \frac{m_z^2}{m_y \cdot m_x}$
Stanton dimensionless number, along directions	$St_x = \frac{Nu_x}{Pe_x} = Nu_x \cdot \frac{1}{Pe_x}$	$\frac{m_x^2}{m_z \cdot m_y} \cdot \frac{m_z \cdot m_y}{m_x^2} = 1^{****}$
	$St_y = \frac{Nu_y}{Pe_y} = Nu_y \cdot \frac{1}{Pe_y}$	$\frac{m_y^2}{m_z \cdot m_x} \cdot 1 = \frac{m_y^2}{m_z \cdot m_x}$
	$St_z = \frac{Nu_z}{Pe_z} = Nu_z \cdot \frac{1}{Pe_z}$	$\frac{m_z^2}{m_x \cdot m_y} \cdot \frac{m_x \cdot m_y}{m_z^2} = 1^{****}$
Fourier dimensionless number, along directions	$Fo_x = \frac{a_x \cdot \tau}{l_x^2} = \frac{a_x \cdot \Delta\tau}{l_x^2} = a_x \cdot \Delta\tau \cdot \frac{1}{l_x^2}$	$\frac{m_y \cdot m_z}{s} \cdot s \cdot \frac{1}{m_x^2} = \frac{m_y \cdot m_z}{m_x^2}$
	$Fo_y = \frac{a_y \cdot \tau}{l_y^2} = \frac{a_y \cdot \Delta\tau}{l_y^2} = a_y \cdot \Delta\tau \cdot \frac{1}{l_y^2}$	$\frac{m_x \cdot m_z}{s} \cdot s \cdot \frac{1}{m_y^2} = \frac{m_x \cdot m_z}{m_y^2}$
	$Fo_z = \frac{a_z \cdot \tau}{l_z^2} = \frac{a_z \cdot \Delta\tau}{l_z^2} = a_z \cdot \Delta\tau \cdot \frac{1}{l_z^2}$	$\frac{m_x \cdot m_y}{s} \cdot s \cdot \frac{1}{m_z^2} = \frac{m_x \cdot m_y}{m_z^2}$

[*]It is considered to be numerically equal to the dimension of work; the work is conventionally considered as a product between a force having the direction along the beam, $F_x \left(N_x = \frac{kg.m_x}{s^2}\right)$ and the displacement along the same direction x *(m_x);[**]The velocity w_0 is normal to the plane where the shear stress is developed; ∇w_0 represents its gradient;[***]When the beam is protected (insulated) by a paint coat, then: $\alpha'_{nf} = \alpha_{ny}$ and $\alpha''_{nf} = \alpha_{nz}$;[****]Therefore it cannot be used in the Dimensional Analysis.*

Table 6.
Variables that can have an influence on the heat transfer in the beam with rectangular or rectangular hole section.

Nomenclature

a	thermal diffusivity (m^2/s);
A	area (m^2);
Bi	Biot number;
c_p	constant-pressure specific heat (W/(m^2 K);
C	heat capacity (J/K);
F	force (N);
F_o	Fourier number;
g	gravitational acceleration (m/s^2);
Gr	Grashof number;
l, L	length (m);
Nu	Nusselt number;
P	perimeter (m);
Pe	Péclet number;
Pr	Prandtl number;
Q	heat (J);
$\dot{Q}$	heat rate (W);

Re	Reynolds number;
St	Stanton number;
t, T	temperature (K);
V	volume (m^3);
w	velocity (m/s);
$S_{\dot{Q}}, S_{L_z}, S_{\Delta t}, S_{\tau}, S_{\lambda_{x\,steel}}, S_{\varsigma}$	scale factor corresponding to the sizes indicated in the index.

Greek symbols

α	convection heat transfer coefficient ($W/(m^2\,K)$);
β	coefficient of volume expansion $(K)^{-1}$;
δ	thickness (m);
Δ	variation;
η	dynamic viscosity (kg/ ms);
λ	thermal conductivity (W/(m K));
ν	kinematic viscosity (m^2/s);
ρ	density (kg/m^3);
ς	shape factor (m^{-1});
τ	time, shear stress (s, N/m^2);
∇	Nabla operator.

Subscripts

x, y, z	directions.

Author details

Ioan Száva*, Teofil-Florin Gălățanu and Ildikó-Renáta Száva
Transilvania University of Brașov, Romania

*Address all correspondence to: apupeter@yahoo.com

References

[1] Carabogdan GI et al. Methods of Analysis of Thermal Energy Processes and Systems. Bucharest, Romania: Technical Publishing House Bucharest; 1989

[2] Baker W et al. Similarity Methods in Engineering Dynamics. Amsterdam: Elsevier Science Publishers; 1991. ISBN-10: 0876715641

[3] Sedov IL. Similarity and Dimensional Methods in Mechanics. Moscow: MIR Publisher; 1982. DOI: 10.1201/9780203739730

[4] Zierep J. Similarity Laws and Modelling. New York: Marcel Dekker; 1971

[5] Buckingham E. On physically similar systems. Physics Magazine. 1914;**4**:345-376. DOI: 10.1103/PhysRev.4.345

[6] Pankhurst RC. Dimensional Analysis and Scale Factor. London, UK: Institute of Physics and the Physical Society by Chapman and Hall; 1964

[7] Bhaskar R, Nigam A. Qualitative physics using dimensional analysis. Artificial Intelligence. 1990;**45**:73-111. DOI: 10.1016/0004-3702(90)90038-2

[8] Langhaar HL. Dimensional Analysis and Theory of Models. New York, NY, USA: John Wiley & Sons Ltd.; 1951

[9] Szekeres P. Mathematical foundations of dimensional analysis and the question of fundamental units. International Journal of Theoretical Physics. 1978;**17**:957-974. DOI: 10.1007/BF00678423

[10] Barr DIH. Consolidation of basics of dimensional analysis. Journal of Engineering Mechanics-Asce. 1984;**110**:1357-1376. DOI: 10.1061/(ASCE)0733-9399(1984)110:9(1357)

[11] Canagaratna SG. Is dimensional analysis the best we have to offer. Journal of Chemical Education. 1993;**70**:40-43

[12] Carinena JF, Santander M. Dimensional analysis. In: Advances in Electronics and Electron Physics. Vol. 72. 1988. pp. 181-258

[13] Carlson DE. Some new results in dimensional analysis. Archive for Rational Mechanics and Analysis. 1978;**68**:191-210

[14] Gibbings JC. A logic of dimensional analysis. Journal of Physics A: Mathematical and General. 1982;**15**:1991-2002. DOI: 10.1088/0305-4470/15/7/011

[15] Coyle RG, Ballicolay B. Concepts and software for dimensional analysis in Modeling. IEEE Transactions on Systems, Man, and Cybernetics. 1984;**14**:478-487. DOI: 10.1109/TSMC.1984.6313242

[16] Martins RDA. The origin of dimensional analysis. Journal of the Franklin Institute. 1981;**311**:331-337

[17] Remillard WJ. Applying dimensional analysis. American Journal of Physics. 1983;**51**:137-140

[18] Romberg G. Contribution to dimensional analysis. Inginieur-Archieve. 1985;**55**:401-412. DOI: 10.1007/BF00537647

[19] Schnittger JR. Dimensional analysis in design. Journal of Vibration, Acoustics, Stress, and Reliability in Design. 1988;**110**:401-407

[20] El Nakla M. On fluid-to-fluid modeling of film boiling heat transfer

using dimensional analysis. International Journal of Multiphase Flow. 2011;**37**(2): 229-234. DOI: 10.1016/j.ijmultiphaseflow.2010.09.004

[21] Szirtes T. The fine art of modelling. SPAR Journal of Engineering and Technology. 1992;**1**:37

[22] Szirtes T. Applied Dimensional Analysis and Modelling. Toronto ON, Canada: McGraw-Hill; 1998

[23] Gálfi BP, Száva I, Şova D, Vlase S. Thermal scaling of transient heat transfer in a round cladded rod with modern dimensional analysis. Mathematics. 1875;**2021**:9. DOI: 10.3390/math9161875

[24] Şova D, Száva RI, Jármai K, Száva I, Vlase S. Modern method to analyze the heat transfer in a symmetric metallic beam with hole. Symmetry. 2022;**14**(4): 769. DOI: 10.3390/sym14040769. Published: 07 April 2022 in Symmetry

[25] Munteanu IR. Investigation concerning temperature field propagation along reduced scale modelled metal structures [PhD thesis]. Brasov, Romania: Transilvania University of Brasov; 2018

[26] Gálfi BP, Száva RI, Száva I, Vlase S, Gălăţanu T, Jármai K, et al. Modern dimensional analysis based on fire-protected steel members' analysis using multiple experiments. Fire. 2022;**5**(6): 210. DOI: 10.3390/fire5060210

Section 2

Engineering Applications of Heat and Mass Transfer

Chapter 5

Heat Transfer within MOSFET Nanotechnology

Oussama Zobiri and Abdelmalek Atia

Abstract

FET (field effect transistor) is the backbone of today's semiconductor electronics. The concept of metal oxide semiconductor field effect transistor (MOSFET) was proposed by Julius Edgar Lilienfeld and the electrical characterization and mathematical theory in Bell Telephone Laboratories. In recent years, the MOSFET miniaturization has been gaining considerable interest. As the size of the device decreases (nanoscale), the heat generated influences the efficiency of the device. In addition to the rapidly increasing semiconductor industries, which have secure significant percentage of the global market, the need to trace back and follow up the progress is quite an essential work which has been carried out time to time. There have been many thermal studies to improve the operation of the device. This chapter reviews and assesses several of the extraction studies currently used to determine the value of temperatures and heat fluxes in nanoelectronics devices versus thermal transfer characteristics.

Keywords: thermal transfer, MOSFET, theory, nanotechnology, device

1. Introduction

Miniaturization of integrated circuits (IC) and the present trend toward nanoscale electronics have led to incredible levels of integration in recent years, with hundreds of millions of semiconductors packed on a chip surface no larger than a few square centimeters. The metal oxide semiconductor field effect transistor (MOSFET) is the most widely used form of insulated gate field effect transistor (FET), found in a wide variety of electronic circuits. Controlling the electrical current flowing through this gadget is crucial [1–3].

The feature size of electrical devices in modern integrated circuits is equal to or less than the mean free path of the phonon, which serves as the foundation for transistors [1, 2, 4, 5]. This reduction in size has an impact on MOS transistors, and the MOSFET channel region is now smaller than the phonon mean free path [6]. The research on thermal transfer within the nano-MOSFET is crucial for demonstrating the devices' thermal stability and efficiency when analyzing integrated circuits made up of transistors [7, 8]. The Joule effect can produce extremely high heat in ultra-small semiconductor devices because the device's operating power cannot be decreased below a specific threshold [9, 10]. Additionally, MOSFET devices' switching times surpass the silicon's phonon relaxation time, which is less than hundred picoseconds [5, 11–16]. As a result of the restricted relaxation period of heat carriers, it is common knowledge that the Fourier law can be significantly altered when applied to the micro- to nanoscale of heat transfer [17]. The Boltzmann transport equation (BTE) and Monte Carlo (MC) solution methods were employed to examine the thermal transfer mechanisms in structures ranging from meso- to macro-scale [18]. The mesoscale method generates statistical data (the particle distribution function) in sub-continuum media resulting from particle movement (vibration), subsequently allowing for the determination of macro-scale physical variables through the aggregation of functional distributions [19–21]. The nano-MOSFET exhibits an additional heat source in addition to the Joule effect. This source, which is represented as a temperature jump boundary condition, originates from the phonon-wall collision at the MOS transistor interface [9, 12–15, 22].

Although most of these investigations based on macroscopic models, many numerical and theoretical predictions of thermal behavior inside nano-MOSFETs have been reported [11–15, 23]. These models appear to require extra terms, such as single and dual phase lag, in order to produce respectable results. Thus, the goal in this work is to evaluate the ability of the proposed mesoscopic thermal model based on the lattice Boltzmann equation to forecast nano-heat transport processes to those of macroscopic models. The advantage of this paradigm is that no new terms are required on the one hand. In contrast, the lattice Boltzmann method (LBM) is thought to be a more efficient method for investigating thermal transfer inside structures than macroscopic methods, particularly in nanostructures [10, 24, 25]. This is because the media is sub-continuum (not continuous) in the presence of phonons (particle vibration) and collisions between them [4, 16, 26–28]. The objective of this study is to evaluate and review a number of the extraction studies that are presently employed to ascertain the value of temperatures and heat fluxes in nanoelectronics devices in relation to thermal transfer characteristics.

2. MOSFET

2.1 A traditional MOSFET

As illustrated in **Figure 1**, a MOSFET device consists of a channel linking the source to the drain, a gate on top of the channel, and a gate oxide (insulator) between the gate and the channel, which prevents the electrical current from the gate to the channel where the gate oxide serves as the dielectric layer so that the gate can sustain as high as MV/cm transverse electric field in order to strongly modulate the conductance of the channel [9, 31].

2.2 Silicon-on-insulator (SOI) MOSFET

It is similar to a regular MOSFET, but a layer of insulator (SiO_2) is added to decrease current leakage from the drain/source junction to the substrate (**Figure 2**) [33]. The insulation layer introduces a lower coupling capacitance from the conducting channel to the substrate compared to typical MOSFETs [34]. It observed that employing SOI MOSFET reduces power dissipation by up to 66% when compared to traditional

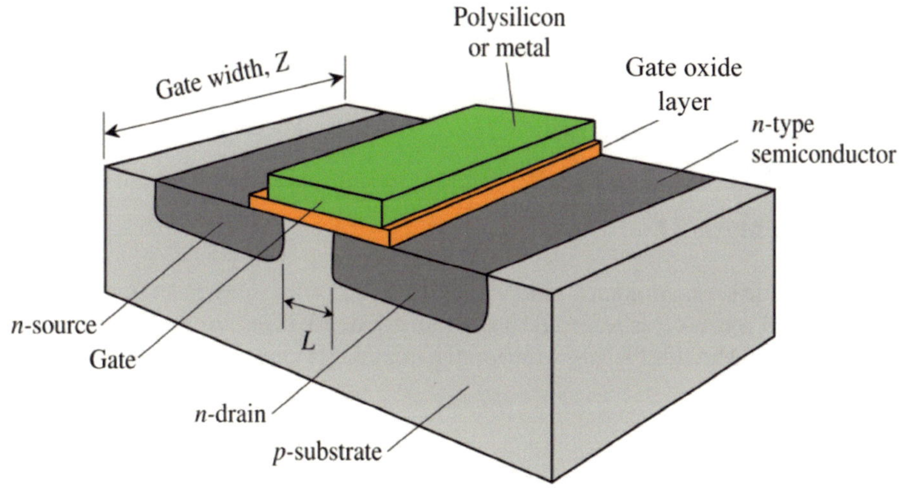

Figure 1.
Conventional MOS device [29, 30].

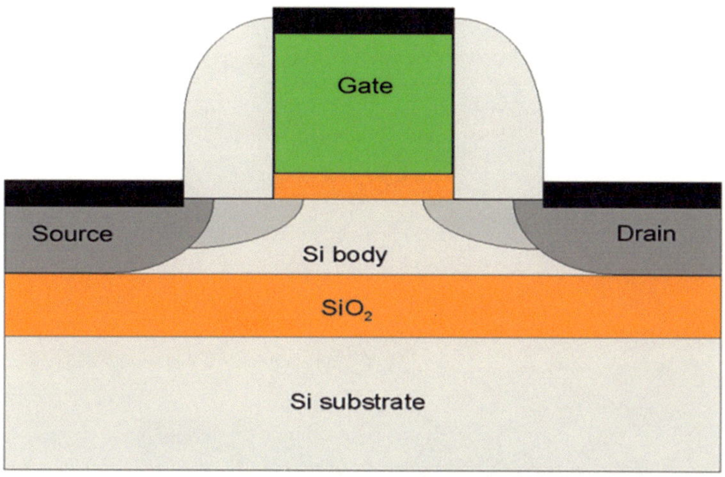

Figure 2.
SOI MOSFET [30, 32].

MOSFETs. And, for the same power dissipation, up to 35% improvement in operating frequency can be obtained [35]. SOI has the advantage of making it easier to isolate electrically a device from the rest of the integrated circuit, resulting in increased packing density. Furthermore, the area of source and drain junctions is greatly reduced, resulting in lower parasitic capacitances. Finally, the depletion width restricted by the thickness of the Si body; thus, it is usually assumed that SOI reduces short channel effects unless source-to-drain coupling through channel and box cannot be ignored. The properties of SOI devices improve as the body thickness reduced. Fully depleted ultra-thin-body SOI (FD UTB SOI) regarded as one of the most effective scaling options [36]. Another advantage of SOI is that it facilitates development of new device concepts [37].

2.3 Double gate MOSFET

These devices include two gates (**Figure 3**) that allow current to pass *via* two channels while concurrently managing the charge in the silicon body layer, as the name implies. Both the front and the back end have two gates. A charge coupling exists between the front and rear gates [38].

2.4 Multi-gate MOSFET

A metal oxide semiconductor field effect transistor with more than two gates, such as three gates (**Figure 4**), is referred to as a multi-gate MOSFET or multi-gate field effect transistor (MuGFET). The additional gates enable control over the charge flow in the channel [39].

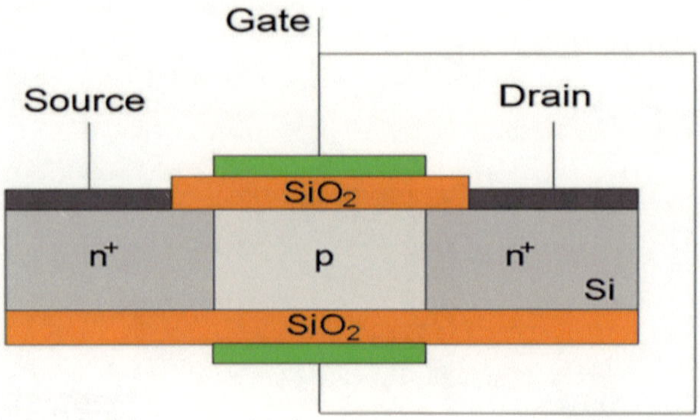

Figure 3.
Double gate transistor [30, 32].

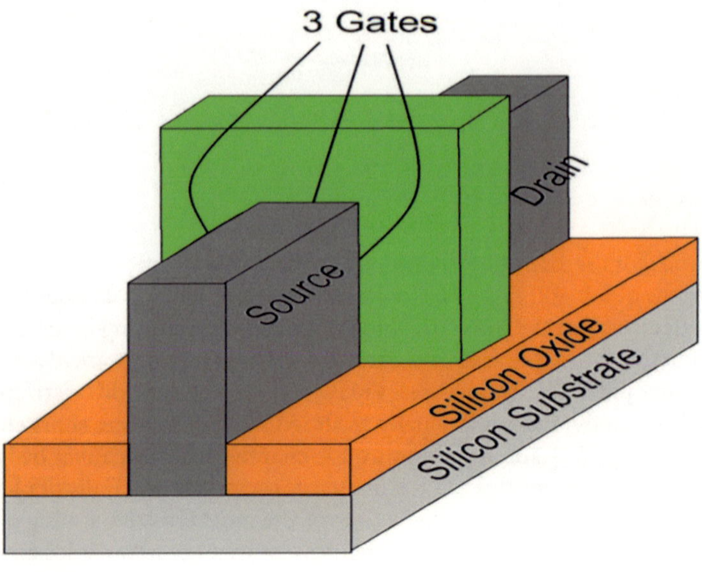

Figure 4.
Tri-Gate MOS [30, 31].

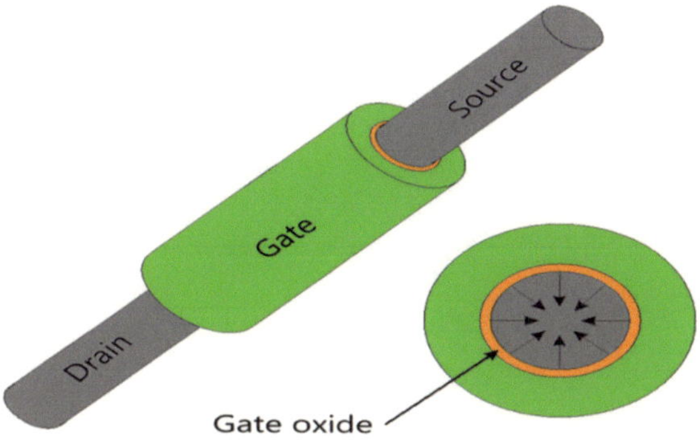

Figure 5.
Surrounding gate MOSFET [30, 32].

2.5 Surrounding gate MOSFET

Controlling the charge flow in the channel is the primary issue with the devices mentioned above. Therefore, a surround gate MOSFET, as illustrated in **Figure 5**, can be used to improve channel control. The channel is completely enclosed by the gate in surround gate devices. It resembles a fin field effect transistor (FinFet) in some way. It offers more accuracy in addition to improved control over the charge flow in the channel [39].

3. Semicoductor materials

Si material: The electronics and solar photovoltaic industries employ high-purity polycrystalline silicon, sometimes known as multicrystalline silicon, and polysilicon, as raw materials [40].

GaAs material: Gallium arsenide (GaAs), a semiconductor with a direct band gap, has a zinc alloy crystal structure. Gallium arsenide is used to make solar cells, laser diodes, infrared light-emitting diodes, microwave-integrated circuits, monolithic microwave-integrated circuits, and optical windows [41].

Ge material: The atomic number of the chemical element germanium is 32, and its symbol is Ge. It is shiny, hard-brittle, grayish-white, and has a silicon-like appearance. It is a metalloid in the carbon group that is chemically comparable to two of its group neighbors, silicon and tin. Like silicon, germanium, and oxygen naturally combine to form complexes [42].

GaN material: Gallium nitride (GaN) is a sensitive semiconductor used in photovoltaic cells that power satellites and radiation-hardened military and space applications. It is useful for power amplification in the microwave spectrum since it can operate at greater temperatures and voltages than GaAs transistors [43]. GaN exhibits great potential and may be found in alloys used in photodiodes, some junction field effect transistors (JFETs), and modulated-doping field effect transistors (MODFETs) devices [43].

4. The reduction of feature size

With hundreds of millions of semiconductors arranged on a chip area no bigger than a few square centimeters and the smallest lateral device feature sizes

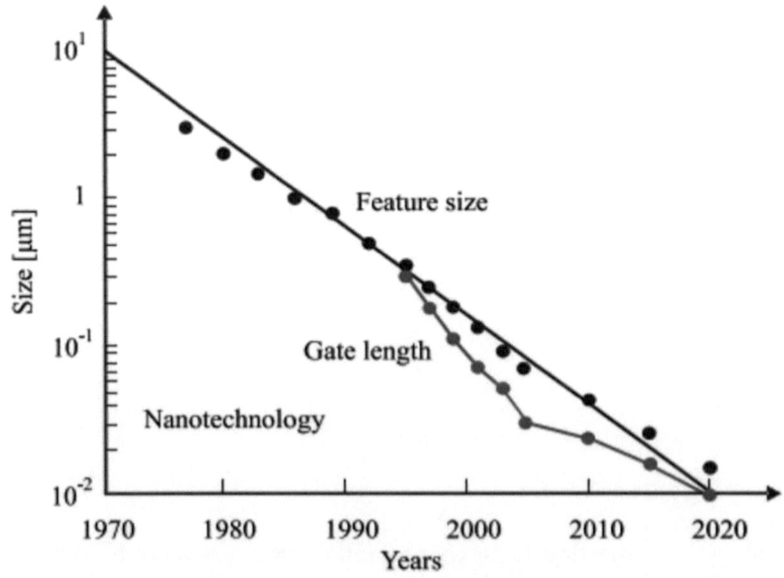

Figure 6.
Feature size as a function of time [44].

approaching 10 nm, electronics have reached the nanoscale [2]. However, complementary MOSFET technology is now used to create over 90% of integrated circuits. The MOSFET has seen innumerable improvements and unimaginable miniaturization. **Figure 6** illustrates how the feature size has decreased almost exponentially [45].

5. Moore's law

Scaling technologies to improve overall performance while consuming less energy and costing less money is what the semiconductor industry wants. Greater speed, greater density, reduced power consumption, reduced form factor, reduced bill of materials, enhanced functionality, and more are some of the metrics that can be used to evaluate the performance of the individual components and the finished chip. In the past, dimensional scaling was enough to generate the performance merits described above; however, this is no longer the case. Tools, processing modules, and material qualities are some of the major barriers to further scaling. Gordon Moore, the Chief Executive Officer CEO of Fairchild Semiconductor and later a co-founder of Intel, used the economics of the integrated circuit to estimate the number of transistors in the chip. He projected in 1965 that the IC would treble annually for the next decade. More transistors imply a lower cost per transistor since "the cost per component is nearly inversely linked to the number of components," according to Moore. He updated his forecast in 1975, pointing out that going forward, the number of transistors will quadruple every 2 years. Moore's law, which was widely used from 1975 until 2015, gives an extremely accurate estimate of how many transistors could fit on a single integrated circuit [45]. Moore's second law predicts that the capital cost of creating ICs will noticeably grow over time, in contrast to Moore's first law, which states that the price of ICs decreases as the number of constituents increases. Costs associated with manufacturing, testing, and research and development rise dramatically with each successive chip generation [46]. In 2015, Gordon Moore asserted, "as investments increase and businesses slow down their rate of innovation, Moore's Law looks to be coming to an end."

However, thanks to advancements like artificial intelligence AI-based devices, application-specific integrated circuits (ASICs), and quantum computing, non-silicon computing is gradually gaining market dominance. Recent advancements in the realm of computing have the potential to result in notable increases in efficiency [47].

6. Heat transport in micro/nanostructure

It makes sense that there are two different approaches to treating heat transport at the micro and nanoscales: theoretical and experimental [48]. There are three categories of theoretical approaches: macroscopic, mesoscopic, and microscopic.

6.1 Microscopic methods

The primary concept of the microscopic approach is numerically solving the basic dynamic equations to directly simulate the motion and interactions of atoms in solids. Newton's equation, which forms the basis of the molecular dynamics simulation approach, was solved within the context of classical mechanics. On the other hand, the Schrödinger equation is resolved using the first-principle approach in the quantum mechanical framework [35, 49, 50].

6.2 Mesoscopic methods

Kinetic modeling based on the phonon Boltzmann equation or its variants, such as the equation of phonon radiative transport, is referred to as mesoscopic approaches. The complex collision term in the Boltzmann equation makes solving it difficult. A full solution and a single mode relaxation time approximation are the two main schools of thinking. Complicated numerical iteration or an *ab initio* variational method are required for the complete answer. Despite the use of empirical parameters, which are frequently modified from experimental data, the single mode relaxation time approximation is more favored due to its simplicity. The phonon Boltzmann equation has been solved using the Monte Carlo method, particularly for heat transport at the micro and nanoscale [16, 51].

6.3 Macroscopic methods

Successful attempts have been made to establish the macroscopic approach for micro- and nanoscale heat transmission concurrently with the advancements of microscopic and mesoscopic methods. The generalized heat transport equations that underpin the macroscopic approach can describe heat transport across a broad range of scales, including the so-called ballistic regime and the so-called phonon hydrodynamic regime, in addition to the traditional diffusive regime where Fourier's law is applicable. The phonon hydrodynamics model is one of the known macroscopic techniques that spreads over a long period [52, 53].

Figure 7 provides a summary of all these techniques. Atomic-scale information is provided by the microscopic approach, statistical information (the particle

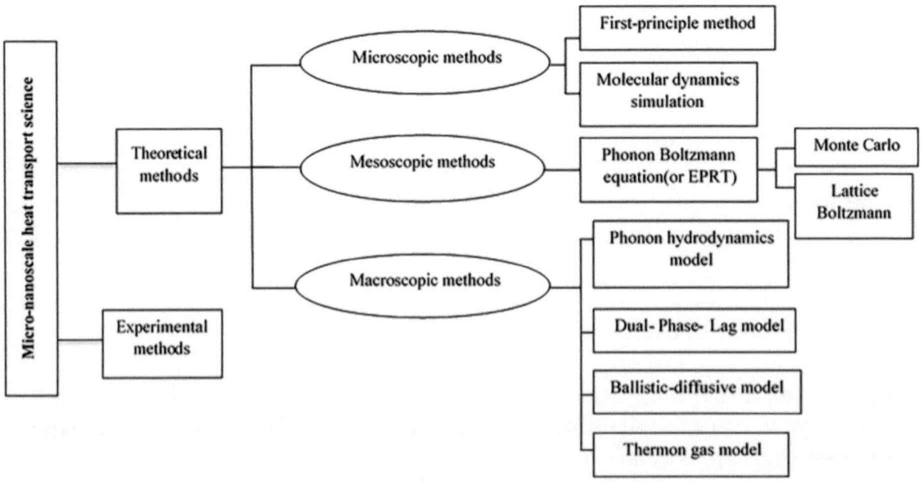

Figure 7.
Current methods in heat transport science [48].

distribution function) is produced by the mesoscopic method, and only a few state variables are used by the macroscopic method for continuum medium [10, 48]. **Table 1** presents the mathematical mascroscale models for heat transport in nano-structure devices.

Where T is the temperature, τ is the relaxation time, k denotes the thermal conductivity, C is volumetric heat capacity, c_p refers to specific heat capacity at pressure constant, and q_v is the volumetric heat generation, whereas T_m is the temperature to the diffusive part τ_t is the relaxation time of temperature, g is the distribution function, g^{eq} is the equilibrium distribution function, ρ is the density, t denotes the time, v is the velocity, and q_b is the ballistic heat flux.

Cattaneo and Vernotti [61, 62] Individually proposed the single phase lag (SPL), a novel concept based on the Fourier law. Because the heat flux vector occurs after the temperature gradient, this equation adds a temporal delay between the two, which is known as the relaxation time. Tzou [63] suggested a novel dual phase lag (DPL)

Model	Formula	References
Ballistic-Diffusive Equation (BDE)	$\tau\frac{\partial^2 T_m}{\partial t^2}+\frac{\partial T_m}{\partial t}=\frac{k}{C}\nabla\nabla T_m-\frac{\nabla q_b}{C}+\frac{q_v}{C}+\frac{\tau}{C}\frac{\partial q_v}{\partial t}$	[54–56]
Single phase lag (SPL)	$\tau\frac{\partial^2 T}{\partial t^2}+\frac{\partial T}{\partial t}=\frac{k}{C}\Delta T+\frac{q_v}{C}+\frac{\tau}{C}\frac{\partial q_v}{\partial t}$	[56, 57]
Dual phase lag (DPL)	$\tau\frac{\partial^2 T}{\partial t^2}+\frac{\partial T}{\partial t}=\frac{k}{C}\text{Ä}T+\frac{q_v}{C}+\frac{\tau}{C}\frac{\partial q_v}{\partial t}+\tau_t\frac{k}{C}\frac{\partial}{\partial t}\Delta T$	[56, 58]
Poisson equation (Fourier law)	$\rho c_p\frac{\partial T}{\partial t}=\nabla.(k\nabla T)+q_v$	[59]
Boltzmann transport equation (BTE)	$\frac{\partial g}{\partial t}+v.\nabla g=-\frac{g-g^{eq}}{\tau}$	[3, 16, 60]

Table 1.
Mathematical models for heat transport in multi-structure device.

model for thermal transfer in nanostructures, in which the temperature gradient may experience the heat flux vector or the heat flux vector may come before the temperature gradient. Chen [64] introduces the ballistic-diffusive equation (BDE), another non-Fourier heat transport model. The Boltzmann transport equation (BTE) served as the basis for this model. The BDE model, according to the study, provides a reliable estimate of the thermal transport in the nanomaterials. Zhang et al. [65] used a 2D MEDICI simulator to examine the temperature distribution in a multi-layered silicon-on-insulator (SOI) MOSFET structure. The findings demonstrated that the new SOI structure could successfully replace the traditional buried dioxide SOI and reduce the self-heating effect. Yang et al. [5] investigated two-dimensional thermal transport in a MOSFET device using the ballistic-diffusive heat conduction equation (BDE). The outcomes of the traditional Boltzmann equation aligned with their findings. Ghazanfarian and Abbassi [66] used the DPL approach with a temperature jump boundary condition (TJBC) to investigate the heat transport in a simplified MOSFET. It was reported that the analytical and numerical results agreed well. Furthermore, they stated that the 2D outcomes might be accurately predicted by integrating the suggested temperature boundary condition type into the DPL approach. Nasri et al. [14] investigated the thermal behavior of a nanoscale MOS transistor device with a temperature jump condition of type using the DPL equation. The application of the temperature jump condition has been shown by the researchers to increase semiconductor-oxide layer collisions at the interface, resulting in an increase in hotspots within the nanodevice. Sinha et al. [67] presented a novel phonon transport technique for non-equilibrium phonon distributions in silicon devices that are nanoscale and about 100 nm in size. They showed a phonon emission spectrum electron-phonon scattering from independent non-Fourier Monte Carlo simulation and managed the isotropic dispersion *via* all phonon locations.

A theoretical method utilizing the Boltzmann transport equation has been developed by Xu and Liqiu [68]. In another study, Xu and Li [69] suggested an analytical BTE model for describing the heat transfer in a MOSFET device. Chen and Weng [70] have shown a slip condition and jump model based on the Beskok–Karniadakis slip law both analytically and numerically and observed that the condition played a major influence in the microchannel's emerging region. Barletta and Zanchini [71] investigated numerically if hyperbolic heat transfer and the local equilibrium hypothesis are compatible using the CV model. Sverdrup et al. [72] used the 2D phonon Boltzmann transport equation (BTE) to study the heat transport in the silicon area of a silicon-on-insulator (SOI) transistor. They discovered that, in comparison to the Fourier's law forecast, the anticipated peak temperature is over 160% higher. Rezgui et al. [54, 55] suggested using a ballistic-diffusive equation to examine the transmission of nano-heat in MOSFETs. According to their research, the effective thermal conductivity decreases as the Knudsen number rises. The findings of this investigation are consistent with the BTE's findings. Belmabrouk et al. [73] used the BDE and Cattaneo-Vernotte models; researchers examined the effect of thermal boundary resistance on thermal transport and found that lowering thermal boundary resistance can improve heat conduction at interfaces. A study [54, 55] investigating the ballistic-diffusive phonon transport along the germanium/silicon interface reported that the temperature jump happens in the interface because of phonon-wall interactions and the surface roughness effects on the interfacial thermal transfer. Analysis of the electrothermal transport involved in the FinFET device was carried out by Rezgui et al. [74]. Lower electron mobility, a high dielectric constant oxide, and lower thermal resistance at the interface could maximize the device's nano-thermal behavior under a stronger electric field.

To reduce the heat generated in the transistor device, other designs have been proposed. Jiang et al. [75] proposed a method for estimating the SiC MOSFET's junction temperature using the dynamic threshold voltage. They experimentally assessed the suggested dynamic threshold voltage calculation circuit using double-pulse testing, and the results showed that the SiC transistor's dynamic threshold voltage changes linearly with junction temperature. Belkhiria et al. [76] suggested using oxide materials with a high dielectric constant in the structure of a partially insulated transistor. The scientists discovered that using an Al_2O_3 layer helps lessen self-heating in partially insulated transistors and that a notable decrease in temperature rise is attained during high-power operation. Ghasemi and Mozaffar [77] created a unique SOI transistor that, using the suggested scheme, was able to completely remove heat. In the buried oxide layer, a T-shaped 4H-SiC area was formed. In comparison to conventional SOI devices, this novel design had a greater cut-off frequency, a shorter delay time, and a higher direct current transconductance. It was also able to absorb heat from the channel active and transfer it to the substrate zone. The vast area of graphene makes it an unacceptable candidate for MOS transistor systems because the zero-band graphene distance reduces the on-off current ratio by 2–20. The current ratio can be raised by forming graphene nanoribbons and shrinking the one-dimensional graphene layer. Graphene layer ribbon's thermal properties are decreasing; however, there are a number of ideas for using this nanomaterial to advance a contemporary [78, 79]. Subrina et al. [80] used the COMSOL Multiphysics software to investigate heat transfer in silicon-on-insulator (SOI) circuits. Both including and excluding graphene's lateral heat spreaders were analyzed. They demonstrated how using graphene or graphene layers with an appropriate heat sink could greatly lower the temperature of the targeted hot spots.

7. Conclusions

The fundamental ideas of nanoscale metal oxide semiconductor field effect transistor (MOSFET) systems are presented in this chapter, along with an explanation of their significance and structure. The equations for analyzing the thermal behavior of these devices were no longer adequate, and it witnessed the amazing advancement in shrinking their size so quickly that it was no longer visible to the human eye. Therefore, additional equation models such as BDE, SPL, and BTE are required to investigate these nanosystems. It briefly discussed the most significant and physically grounded semi-classical and quantum transport methods at the moment. It is noteworthy that device simulation has attained a far greater level of maturity than process simulation because of these advancements. Actually, the fundamental physics up to the ballistic transport regime and the point at which quantum interference effects begin to dominate device behavior may be captured by particle-based device simulators. But in the end, both the engineering-oriented transport simulations and the complex physics-oriented electronic structure computations must be included in the modeling of nanoelectronic devices. Recently, there have been many scientific debates over transport theory, basis representation, and the actual use of a simulator that can describe a realistic device.

Acknowledgements

The current research is supported by the DZ General Directorate of Scientific Research and Technological Development.

Conflict of interest

We wish to confirm that there are no known conflicts of interest associated with this publication.

Nomenclature

ASICs	application-specific integrated circuits
BTE	Boltzmann transport equation
CEO	chief executive officer
DG	double gate
FD UTB	fully depleted ultra-thin body
FET	field effect transistor
GaAs	gallium arsenide
GaN	gallium nitride
IC	integrated circuit
LBM	lattice Boltzmann method
MC	Monte Carlo
MOSFET	metal oxide semiconductor field effect transistor
SOI	silicon on insulator
T	temperature, K
τ	relaxation time, s
k	thermal conductivity, W/m.K
C	volumetric heat capacity, $J/m^3.K$
c_p	specific heat capacity at pressure constant, J/kg.K
q_v	volumetric heat generation, W/m^3
T_m	diffusive temperature, K
τ_t	relaxation time of temperature, s
g	distribution function
g^{eq}	equilibrium distribution function
ρ	density, kg/m^3
t	time, s
v	velocity, m/s
q_b	ballistic heat flux, W/m^2

Author details

Oussama Zobiri* and Abdelmalek Atia
LEVRES Lab, Faculty of Technology, University of El Oued, El Oued, Algeria

*Address all correspondence to: zobiri-oussama@univ-eloued.dz

References

[1] Łukasiak L, Jakubowski A. History of semiconductors. Journal of Telecommunications and Information Technology. 2010:3-9

[2] Pop E, Sinha S, Goodson KE. Heat generation and transport in nanometer-scale transistors. Proceedings of the IEEE. 2006;**94**(8):1587-1601. DOI: 10.1109/JPROC.2006.879794

[3] Oussama Z et al. Study of robin condition influence on phonon nano-heat conduction using meso-scale method in MOSFET and SOI-MOSFET devices. Materials Today Communications. 2021;**26**:102031. DOI: 10.1016/j.mtcomm.2021.102031

[4] Chen G. Phonon heat conduction in low-dimensional structures. Semiconductors and Semimetals. 2001a;**71**:203-259. DOI: 10.1016/S0080-8784(01)80130-7

[5] Yang R, Chen G, Laroche M, Taur Y. Simulation of nanoscale multidimensional transient heat conduction problems using ballistic-diffusive equations and phonon Boltzmann equation. Journal of Heat Transfer. 2005;**127**(3):298-306. DOI: 10.1115/1.1857941

[6] Taur Y, Wann CH, Frank DJ. 25 nm CMOS design considerations. In: Paper Presented at the International Electron Devices Meeting 1998. Technical Digest (Cat. No. 98CH36217); 1998

[7] Aljawarneh S. Cloud security engineering: Avoiding security threats the right way. In: Cloud Computing Advancements in Design, Implementation, and Technologies. IGI Global; 2013. pp. 147-153

[8] Aljawarneh S, Aldwairi M, Yassein MB. Anomaly-based intrusion detection system through feature selection analysis and building hybrid efficient model. Journal of Computational Science. 2018;**25**:152-160. DOI: 10.1016/j.jocs.2017.03.006

[9] Mansouri K, Zobiri O, Atia A, Arıcı M. A python implementation based lattice Boltzmann method for thermal behavior analysis in silicon carbide MOSFET. Micro and Nanostructures. 2024;**187**:207769. DOI: 10.1016/j.micrna.2024.207769

[10] Oussama Z. Effect of specularity parameter and convective coefficient on heat transport in semiconductor devices based on mesoscopic method. Physica B: Condensed Matter. 2023b;**670**:415363. DOI: 10.1016/j.physb.2023.415363

[11] Ghazanfarian J, Shomali Z. Investigation of dual-phase-lag heat conduction model in a nanoscale metal-oxide-semiconductor field-effect transistor. International Journal of Heat and Mass Transfer. 2012;**55**(21-22):6231-6237. DOI: 10.1016/j.ijheatmasstransfer.2012.06.052

[12] Nasri F, Ben Aissa MF, Belmabrouk H. Effect of second-order temperature jump in metal-oxide-semiconductor field effect transistor with dual-phase-lag model. Microelectronics Journal. 2015c;**46**(1):67-74. DOI: 10.1016/j.mejo.2014.10.007

[13] Nasri F, Ben Aissa MF, Belmabrouk H. Microscale thermal conduction based on Cattaneo-Vernotte model in silicon on insulator and double gate MOSFETs. Applied Thermal Engineering.

2015d;**76**:206-211. DOI: 10.1016/j.applthermaleng.2014.11.038

[14] Nasri F, Echouchene F, Ben Aissa MF, Graur I, Belmabrouk H. Investigation of self-heating effects in a 10-nm SOI-MOSFET with an insulator region using electrothermal modeling. IEEE Transactions on Electron Devices. 2015a;**62**(8):2410-2415. DOI: 10.1109/TED.2015.2447212

[15] Nasri F, Aissa MFB, Gazzah MH, Belmabrouk H. 3D thermal conduction in a nanoscale tri-gate MOSFET based on single-phase-lag model. Applied Thermal Engineering. 2015b;**91**:647-653. DOI: 10.1016/j.applthermaleng.2015.08.045

[16] Zobiri O, Atia A, Arıcı M. Analysis of heat conduction in a nanoscale metal oxide semiconductor field effect transistor using lattice Boltzmann method. Energy Sources, Part A: Recovery, Utilization, and Environmental Effects. 2020;**43**:1-15. DOI: 10.1080/15567036.2020.1786194

[17] Tien CL, Majumdar A, Gerner FM. Transport Microscale Energy. Washington, DC: Taylor & Francis; 1998

[18] Oussama Z. A three-dimensional analysis of heat transfer based on mesoscopic method in nanoscale Si-MOSFET and Gr-FET. Micro and Nanostructures. 2022;**163**:107123. DOI: 10.1016/j.spmi.2021.107123

[19] Lewis EE, Miller WF. Computational Methods of Neutron Transport. New York, NY: John Wiley and Sons, Inc; 1984

[20] Modest MF. Radiative Heat Transfer. New York: Academic Press; 2003

[21] Tellier CR, Tosser AJ. Size Effects in Thin Films. Vol. 2. Elsevier; 2016

[22] Nasri F, Atri M. Channel length effect on heat transfer process in nano MOSFETs. In: Paper Presented at the 2017 International Conference on Engineering & MIS (ICEMIS). 2017

[23] Nasri F, Aissa MFB, Belmabrouk H. Nonlinear electrothermal model for investigation of heat transfer process in a 22-nm FD-SOI MOSFET. IEEE Transactions on Electron Devices. 2017;**64**(4):1461-1466. DOI: 10.1109/TED.2017.2666262

[24] Mohamad AA. Lattice Boltzmann Method: Fundamentals and Engineering Applications with Computer Codes. Springer Science & Business Media; 2011

[25] Nabovati A, Sellan DP, Amon CH. On the lattice Boltzmann method for phonon transport. Journal of Computational Physics. 2011;**230**(15):5864-5876. DOI: 10.1016/j.jcp.2011.03.061

[26] Chen G. Nanoscale Energy Transport and Conversion: A Parallel Treatment of Electrons, Molecules, Phonons, and Photons. Oxford University Press; 2005

[27] Mahabadian MA, Ghayyem MA, Jamialahmadi M. Multicomponent multiphase lattice-Boltzmann modeling of fingering during immiscible displacement. Energy Sources, Part A: Recovery, Utilization, and Environmental Effects. 2015;**37**(6):642-648. DOI: 10.1080/15567036.2011.585377

[28] Mahabadian MA, Jamialahmadi M. The investigation of longitudinal dispersion coefficient in a miscible displacement process using multicomponent multiphase Shan-Chen lattice-Boltzmann modeling. Energy Sources, Part A: Recovery, Utilization, and Environmental Effects. 2012;**34**(24):2268-2279. DOI: 10.1080/15567036.2011.563271

[29] Mishra U, Singh J. Semiconductor Device Physics and Design. Dordrecht: Springer; 2008

[30] Zobiri O. Contribution Study of Meso-Scale Modelling and Simulation of Heat Transfer Phenomenon in Semiconductor Systems. Algeria: University of Eloued جامعة الشهيد حمه لخضر; 2022

[31] Samian RS, Abbassi A, Ghazanfarian J. Thermal investigation of common 2D FETs and new generation of 3D FETs using Boltzmann transport equation in nanoscale. International Journal of Modern Physics C. 2013;**24**(9):1350064. DOI: 10.1142/S0129183113500642

[32] Jakubowski A, Łukasiak L. CMOS evolution. Development limits. Materials Science Poland. 2008;**26**(1):5-20

[33] Woo JCS, Terrill KW, Vasudev PK. Two-dimensional analytic modeling of very thin SOI MOSFETs. IEEE Transactions on Electron Devices. 1990;**37**(9):1999-2006

[34] Chang L, Choi Y-k, Ha D, Ranade P, Xiong S, Bokor J, et al. Extremely scaled silicon nano-CMOS devices. Proceedings of the IEEE. 2003;**91**(11):1860-1873

[35] DasGupta N, Dasgupta A. Semiconductor Devices: Modelling and Technology. PHI Learning Pvt. Ltd.; 2004

[36] Skotnicki T, Hutchby JA, King T-J, Wong HSP, Boeuf F. The end of CMOS scaling: Toward the introduction of new materials and structural changes to improve MOSFET performance. IEEE Circuits and Devices Magazine. 2005;**21**(1):16-26. DOI: 10.1109/MCD.2005.1388765

[37] Colinge J-P. Multiple-gate soi mosfets. Solid-State Electronics. 2004;**48**(6):897-905. DOI: 10.1016/j.sse.2003.12.020

[38] Lim H-K, Fossum JG. Threshold voltage of thin-film silicon-on-insulator (SOI) MOSFET's. IEEE Transactions on Electron Devices. 1983;**30**(10):1244-1251

[39] Ashaf A, Tyagi M, Mani P. To study high performance analysis of surround gate SOI MOSFET. International Journal of Engineering & Technology. 2018;**7**(2.8):191-194

[40] Nabil M, Elnouby M, Al-Askar AA, Kowalczewski PŁ, Abdelkhalek A, Behiry SI. Porous silicon nanostructures: Synthesis, characterization, and their antifungal activity. Open Chemistry. 2024;**22**(1):20230169. DOI: 10.1515/chem-2023-0169

[41] Moss SJ, Ledwith A. Chemistry of the Semiconductor Industry. Springer Science & Business Media; 1989

[42] Emsley J. Nature's Building Blocks: An AZ Guide to the Elements. USA: Oxford University Press; 2011

[43] Lidow A, Brandon Witcher J, Smalley K. Enhancement mode gallium nitride (eGaNTM) FET characteristics under long term stress. In: Paper Presented at the Proc. GOMAC Tech. 2011

[44] International Technology Roadmap for Semiconductors. 2008." Available from: http://www.itrs.net/Links/2008ITRS/home2008.htm

[45] Arsov GL. Celebrating 65th anniversary of the transistor. Electronics. 2013;**17**(2):63-70. DOI: 10.7251/ELS1317063A

[46] Moore GE. Progress in digital integrated electronics. In: Paper

Presented at the Electron Devices Meeting. 1975

[47] Components Hardware. Intel Processor History. 2024. Available from: http://www.interfacebus.com/intel-processor-types-release-date.html

[48] Guo Y, Wang M. Phonon hydrodynamics and its applications in nanoscale heat transport. Physics Reports. 2015;**595**:1-44. DOI: 10.1016/j.physrep.2015.07.003

[49] Yacobi BG. Semiconductor Materials: An Introduction to Basic Principles. Springer Science & Business Media; 2003

[50] Ziman JM. Electrons and Phonons: The Theory of Transport Phenomena in Solids. Oxford University Press; 2001

[51] Malik NR. Electronic Circuits: Analysis, Simulation, and Design. Prentice-Hall, Inc; 1995

[52] Kittel C. Introduction to Solid State Physics. Vol. 8. New York: Wiley; 1976

[53] Nishizawa J-I. Junction field-effect devices. In: Semiconductor Devices for Power Conditioning. Springer; 1982. pp. 241-272

[54] Rezgui H, Faouzi Nasri MF, Aissa B, Guizani AA, Ravariu C. Study of heat dissipation mechanism in nanoscale MOSFETs using BDE model. In: Green Electronics. London, UK, London, UK: IntechOpen; 2018a. p. 15

[55] Rezgui H, Nasri F, Aissa MFB, Blaabjerg F, Belmabrouk H, Guizani AA. Investigation of heat transport across Ge/Si interface using an enhanced ballistic-diffusive model. Superlattices and Microstructures. 2018b;**124**:218-230. DOI: 10.1016/j.spmi.2018.09.018

[56] Moghaddam M, Ghazanfarian J, Abbassi A. Implementation of DPL-DD model for the simulation of nanoscale MOS devices. IEEE Transactions on Electron Devices. 2014;**61**(9):3131-3138. DOI: 10.1109/TED.2014.2342037

[57] Belkhiria M, Echouchene F, Jaba N, Bajahzar A, Belmabrouk H. 2-D-nonlinear electrothermal model for investigating the self-heating effect in GAAFET transistors. Review of IEEE Transactions on Electron Devices. 2021;**68**(3):954-961. DOI: 10.1109/TED.2020.3048919

[58] Nasri F, Aissa MFB. Thermal analysis of phonon temperature in nano transistor based on DPL model coupled with temporal temperature jump. In: Paper Presented at the 2017 International Conference on Engineering & MIS (ICEMIS). 2017

[59] Both S, Czél B, Fülöp T, Gróf G, Gyenis Á, Kovács R, et al. Deviation from the Fourier law in room-temperature heat pulse experiments. Journal of Non-Equilibrium Thermodynamics. 2016;**41**(1):41-48. DOI: 10.1515/jnet-2015-0035

[60] Oussama Z. Analysis of heat conduction in a nanoscale metal oxide semiconductor field effect transistor using lattice Boltzmann method. Energy Sources, Part A: Recovery, Utilization, and Environmental Effects. 2023a;**45**(3):8864-8878. DOI: 10.1080/15567036.2020.1786194

[61] Cattaneo C. A form of heat-conduction equations which eliminates the paradox of instantaneous propagation. Comptes Rendus. 1958;**247**:431

[62] Vernotte P. Les paradoxes de la theorie continue de l'equation de la chaleur. Comptes Rendus. 1958;**246**:3154-3155

[63] Tzou DY. A unified field approach for heat conduction from macro-to micro-scales. Journal of Heat Transfer. 1995;**117**:8-16. DOI: 10.1115/1.2822329

[64] Chen G. Ballistic-diffusive heat-conduction equations. Physical Review Letters. 2001b;**86**(11):2297. DOI: 10.1103/PhysRevLett.86.2297

[65] Zhang ZX, Lin Q, Zhu M, Lin CL. A new structure of SOI MOSFET for reducing self-heating effect. Ceramics International. 2004;**30**(7):1289-1293. DOI: 10.1016/j.ceramint.2003.12.033

[66] Ghazanfarian J, Abbassi A. Investigation of 2D transient heat transfer under the effect of dual-phase-lag model in a nanoscale geometry. International Journal of Thermophysics. 2012;**33**(3):552-566. DOI: 10.1007/s10765-012-1164-6

[67] Sinha S, Pop E, Dutton RW, Goodson KE. Non-equilibrium phonon distributions in sub-100 nm silicon transistors. Journal of Heat Transfer. 2006;**128**(7):638-647. DOI: 10.1115/1.2194041

[68] Xu M, Wang L. Dual-phase-lagging heat conduction based on Boltzmann transport equation. International Journal of Heat and Mass Transfer. 2005;**48**(25-26):5616-5624. DOI: 10.1016/j.ijheatmasstransfer.2005.05.040

[69] Xu M, Li X. The modeling of nanoscale heat conduction by Boltzmann transport equation. International Journal of Heat and Mass Transfer. 2012;**55**(7-8):1905-1910. DOI: 10.1016/j.ijheatmasstransfer.2011.11.045

[70] Weng HC. Drag reduction and heat transfer enhancement over a heated wall of a vertical annular microchannel. International Journal of Heat and Mass Transfer. 2009;**52**(3-4):1075-1079. DOI: 10.1016/j.ijheatmasstransfer.2008.06.022

[71] Barletta A, Zanchini E. Hyperbolic heat conduction and local equilibrium: A second law analysis. International Journal of Heat and Mass Transfer. 1997;**40**(5):1007-1016. DOI: 10.1016/0017-9310(96)00211-6

[72] Sverdrup PG, Sungtaek Ju Y, Goodson KE. Sub-continuum simulations of heat conduction in silicon-on-insulator transistors. Journal of Heat Transfer. 2001;**123**(1):130-137. DOI: 10.1115/1.1337651

[73] Belmabrouk H, Rezgui H, Nasri F, Aissa MFB, Guizani AA. Interfacial heat transport across multilayer nanofilms in ballistic–diffusive regime. The European Physical Journal Plus. 2020;**135**(1):109. DOI: 10.1140/epjp/s13360-020-00180-7

[74] Rezgui H, Nasri F, Nastasi G, Aissa MFB, Rahmouni S, Romano V, et al. Design optimization of nanoscale electrothermal transport in 10 nm SOI FinFET technology node. Journal of Physics D: Applied Physics. 2020;**53**(49):495103. DOI: 10.1088/1361-6463/abaf7c

[75] Jiang X, Wang J, Hengyu Y, Chen J, Zeng Z, Yang X, et al. Online junction temperature measurement for SiC MOSFET based on dynamic threshold voltage extraction. IEEE Transactions on Power Electronics. 2020;**36**(4):3757-3768. DOI: 10.1109/TPEL.2020.3022390

[76] Belkhiria M, Echouchene F, Jaba N, Bajahzar A, Belmabrouk H. Impact of high-k gate dielectric on self-heating effects in PiFETs structure. Review of IEEE Transactions on Electron Devices. 2020;**67**(9):3522-3529. DOI: 10.1109/TED.2020.3012418

[77] Ghasemi H, Mozaffari MH. A simple proposal to reduce self-heating effect in SOI MOSFETs. Silicon. 2020;**13**(4189):4189-4198. DOI: 10.1007/s12633-020-00773-y

[78] Fiori G, Iannaccone G. Simulation of graphene nanoribbon field-effect transistors. IEEE Electron Device Letters. 2007;**28**(8):760-762. DOI: 10.1109/LED.2007.901680

[79] Son JG, Son M, Moon K-J, Lee BH, Myoung J-M, Strano MS, et al. Sub-10 nm graphene nanoribbon array field-effect transistors fabricated by block copolymer lithography. Advanced Materials. 2013;**25**(34):4723-4728. DOI: 10.1002/adma.201300813

[80] Subrina S, Kotchetkov D, Balandin AA. Heat removal in silicon-on-insulator integrated circuits with graphene lateral heat spreaders. IEEE Electron Device Letters. 2009;**30**(12):1281-1283. DOI: 10.1109/LED.2009.2034116

Chapter 6

Thermal Energy Conversion in Wearable Thermoelectric Generators

Diana Enescu

Abstract

Wearable thermoelectric generators (w-TEGs) convert body heat into electrical energy, providing a continuous power source for small portable wearable electronics. In the context of energy harvesting, w-TEGs contribute to providing usable electrical energy from a source independent of external networks. The w-TEGs provide this energy directly to small portable electronic devices, allowing them to operate without needing external batteries or frequent recharging. This chapter discusses the principles and advancements in w-TEGs, focusing on their ability to convert body heat into electrical energy. The main topics include the mechanisms of heat transfer—conduction, convection, radiation, and evaporation—within the human body and their role in improving w-TEG performance. The chapter also includes aspects related to design improvements, material innovations, and thermal management strategies to increase energy conversion efficiency, and user comfort.

Keywords: wearable thermoelectric generators, energy harvesting, thermal resistance, body heat conversion, heat transfer

1. Introduction

The human body has significant potential as an energy source, offering various possibilities for energy harvesting, depending on mechanical aspects (linked to the body movement) and thermal aspects (depending on the temperature of the body and the surroundings).

The principle of harvesting energy from human body movement is based on the fact that the amount of energy used by the body per day is 10.7 MJ [1]. The body produces this energy by metabolising nutrients from food (proteins, carbohydrates and fats). This amount is comparable to the energy stored in approximately 800 AA batteries (2500 mAh each), weighing about 20 kg [2, 3]. The energy content of food is generally 35–100 times higher than that of most batteries currently in use, depending on the type of battery [4]. This means that most of the energy from food is released into the environment as heat.

On the thermal side, wearable thermoelectric generators (w-TEGs) are devices that can harvest and convert heat into electrical energy. The w-TEGs convert waste heat dissipated from the human body into electrical energy through the Seebeck effect, a thermoelectric phenomenon where a temperature difference between two materials generates voltage. The generated power comes from the difference in temperature between the human body, typically at 37°C, and the temperature of the air immediately surrounding the body [5].

The basic principle is based on the fact that the human body, acting as a constant heat source, can serve as an energy source when paired with TEG devices.

The theoretical maximum efficiency of any system that converts body heat into energy, such as a TEG, can be estimated using Carnot efficiency. The Carnot efficiency estimates the theoretical upper limit of how efficiently any system can convert heat into another form of energy, including electrical energy. In this case, the Carnot efficiency related to the body provides the upper limit of efficiency based on the temperature difference between a heat source (the human body) and a cold source (the surrounding environment), being calculated as follows:

$$\eta_{\mathrm{C\ body}} = \frac{T_{\mathrm{body}} - T_{\mathrm{surr}}}{T_{\mathrm{surr}}} \tag{1}$$

The Carnot efficiency at a body temperature of 310 K (37°C) and the surrounding environment temperature of 273 K (0°C) is 12% [2, 6, 7]. The Carnot efficiency refers to the maximum amount of heat energy from the body that could theoretically be converted into useful work under perfect conditions. It shows the limit beyond which no energy conversion device can go.

The TEG devices operate far below the Carnot limit, and their performance is determined by the properties of the thermoelectric material, described by the dimensionless figure of merit, $Z\overline{T}$. This parameter depends on three key material properties (Seebeck coefficient α, electrical conductivity σ, and thermal conductivity k), along with the mean absolute temperature $\overline{T}$. The equation for $Z\overline{T}$ is expressed as:

$$Z\overline{T} = \frac{\alpha^2 \cdot \sigma \cdot \overline{T}}{k} \tag{2}$$

where the mean absolute temperature $\overline{T}$ is calculated as the arithmetic mean between the temperature at the hot side T_{h} and the temperature on the cold side T_{c}. The term $\alpha^2 \cdot \sigma$, known as the power factor, is the main indicator of the thermoelectric material performance. For optimal performance, a thermoelectric material needs to have high power factor (which means that both α and σ are high) while maintaining low thermal conductivity k.

The maximum conversion efficiency of the TEG depends on $Z\overline{T}$ and the temperature difference across the thermoelectric elements, $\Delta T = T_{\mathrm{h}} - T_{\mathrm{c}}$.

$$\eta_{\max} = \left(\frac{\Delta T}{T_{\mathrm{h}}}\right) \cdot \frac{\sqrt{1+Z\overline{T}} - 1}{\sqrt{1+Z\overline{T}} + \frac{T_{\mathrm{c}}}{T_{\mathrm{h}}}} = \eta_{\mathrm{C\ TEG}} \cdot \frac{\sqrt{1+Z\overline{T}} - 1}{\sqrt{1+Z\overline{T}} + \frac{T_{\mathrm{c}}}{T_{\mathrm{h}}}} \tag{3}$$

where $\eta_{\mathrm{C\ TEG}}$ is the Carnot efficiency related to the TEG device, serving as a limit on utilising waste heat for thermoelectric power generation [8]. The maximum theoretical efficiency of a TEG is directly affected by the temperature difference ΔT.

For converting body heat into electricity, the most effective thermoelectric materials are bismuth-based alloys combined with antimony, tellurium, or selenium [9].

Typically, $Z\overline{T} \approx 1$ for these materials. Future advancements in materials could improve TEG efficiency. The efficiency of TEGs depends on the temperature difference between the body and the surrounding environment. The greater the temperature difference, the higher the conversion efficiency, and similarly, a smaller temperature difference results in lower conversion efficiency.

Another important aspect of energy harvesting from the human body is understanding how heat is lost to the environment. There are two main ways this happens: through direct heat transfer (sensible heat) and through evaporation (sweating, which is latent heat). Sensible heat, also known as heat transfer caused by temperature differences, refers to the exchange of thermal energy due to temperature changes [10, 11]. Latent heat, often described as the transfer of heat during phase changes, involves the transfer of thermal energy when a substance changes its state. This includes heat absorption during vaporisation and heat release during condensation [12]. However, TEG devices can only exploit the temperature difference related to sensible heat, meaning that the heat lost through evaporation is wasted [2, 3]. In a stationary state, about 60% of the body's heat is lost through sensible heat transfer, while 40% is released as latent heat via water vapour. When the body is in motion, these proportions change, with around 35% of the heat lost through sensible heat and 65% through latent heat transfer due to increased sweating [3]. Under normal human activity, the values of the sensible heat dissipation undergone by the human body are in the range of 50–150 W/m^2 [13].

Thermoelectric conversion devices operate within a maximum Carnot efficiency range of 3.3–5.5% [14, 15]. For instance, when a person at rest generates 100 W of total power, only 1.98–3.3 W of sensible heat can be converted to electricity, while 1.28 to 2.2 W of latent heat from evaporated moisture can be harnessed (as shown in **Table 1**) [2, 3]. This highlights the importance of utilising both sensible and latent heat simultaneously to optimise the TEG performance.

The wearable, flexible thermoelectric devices designed for skin heat harvesting are divided into two categories: w-TEGs [16–18], which harvest sensible heat, and wearable moist-electric generators (w-MEGs), which harness latent heat from moisture [19, 20]. In the context of wearable devices, "harvesting" energy from body heat

Activity	Total generated power, W	Sensible Heat output, W	Latent Heat output, W	Energy conversion efficiency range, %	Generated power (Sensible), W	Generated power (Latent), W
Seated at rest	100	60	40	3.3–5.5	1.98–3.3	1.32–2.2
Seated light work	120	65	55	3.3–5.5	2.15–3.58	1.82–3.02
Seated eating	170	75	95	3.3–5.5	2.48–4.12	3.14–5.22
Walking at 4.83 km/h	305	100	205	3.3–5.5	3.3–5.5	6.77–11.28
Heavy work (lifting)	465	165	300	3.3–5.5	5.45–9.07	9.9–16.5
Athletics	525	185	340	3.3–5.5	6.11–10.18	11.22–18.7

Table 1.
The maximum theoretical output power from both sensible and latent heat for different activities (Adapted from [2, 3]).

refers to collecting energy for later use, while "harnessing" latent heat from moisture implies directly using that energy as it is produced.

The wearable TEGs are particularly attractive because they offer portable, noiseless, and sustainable energy generation. They work best when there is an optimal temperature gradient between the body and the environment. By improving heat transfer, the efficiency of energy conversion can be significantly enhanced, allowing w-TEGs to power small devices like fitness trackers, medical sensors, and other wearable electronics. The next section discusses the heat transfer principles in the human body, highlighting their role in thermoregulation. Understanding these principles is essential for optimising the performance of w-TEGs. Various physiological mechanisms contribute to this process, ensuring continuous heat exchange that can be used for energy generation.

2. Heat transfer principles in the human body

Heat transfer from the human body to the environment is important for thermoregulation. There are various physiological mechanisms which are coordinated together. These mechanisms are:

- Insulation, which keeps a comfortable core temperature of the body, slowing down the heat transfer from the body to the environment.

- Sweating, which leads to water evaporation and due to this phenomenon, the body is cooled.

- Shivering, which produces heat through the contraction of the muscle.

- Vasodilation and vasoconstriction, which distribute heat throughout the body and regulate blood flow.

The physiological mechanisms are the processes the body uses to regulate its temperature. The human body keeps a core temperature of about 37°C, while the room temperature range varies from 0–35°C [3]. When the body's core temperature decreases, the thermoregulation system activates physiological mechanisms like vasoconstriction and shivering, which are useful to warm the body. Conversely, when the body core temperature increases, the thermoregulation system activates physiological mechanisms like vasodilation and sweating, which are useful to cool the body [21]. In addition to these thermoregulatory mechanisms, the human body continuously generates heat through metabolic processes—the chemical reactions occurring within the body's cells. The intense chemical processes in the cells that help regulate the body's temperature at about 37°C while supporting essential body functions are collectively known as metabolism. Metabolic heat is important for maintaining a stable internal temperature of around 37°C, which is necessary for the body to function properly. The ability of the body to keep this constant temperature, even when the environment changes, is called homeothermy. Homeothermy is closely linked to the metabolic rate, as the energy produced by the body through metabolism helps maintain this steady temperature [22]. The metabolic rate, which is measured in metabolic equivalents (Met), refers to the amount of energy the body uses. To better understand how the body manages this temperature, it is important to understand the main mechanisms through which heat is transferred to the environment, such as conduction, convection, radiation, and evaporation.

2.1 Conductive heat transfer

Conductive heat transfer occurs when part of the body comes into direct contact with a solid surface. The amount of heat transferred depends on several factors, including the surface area in contact and the thermal properties of the solid material. In steady-state conditions, the heat transfer can be estimated using the material's thermal conductivity and the temperature difference between the skin and the solid surface. The rate of heat transfer through conduction can be expressed as:

$$\dot{Q}_{cond} = k \cdot S \cdot (T_{skin} - T_{surface}) \tag{4}$$

where k is the thermal conductivity in W/(m·K), S is the contact surface area between the skin body and the solid surface, measured in m^2, T_{skin} is the temperature of the skin body in K, and $T_{surface}$ is the temperature of the solid surface in K.

In dynamic conditions, such as when the body touches surfaces with different thermal capacities, heat transfer can be influenced by what is known as "thermal inertia". Thermal inertia describes how quickly a material can absorb or release heat when its temperature changes. Materials with higher thermal inertia take longer to change temperature and, therefore, will affect the skin temperature more slowly upon contact. Depending on the size of the contact area and the thermal properties of the material, conductive heat transfer can vary significantly, being greater in situations where there is extended contact and the material has high thermal inertia, such as when a person is lying down [7, 23, 24].

The human body releases the heat produced by metabolism through the skin and lungs. This happens in two main ways: sensible heat (through convection and radiation) and latent heat (through evaporation), as shown in **Figure 1**.

Sensible heat is the heat that causes a temperature change, which can be felt or measured but does not involve a phase change. Sensible heat is transferred from the skin to the environment and from the lungs to the environment through convection

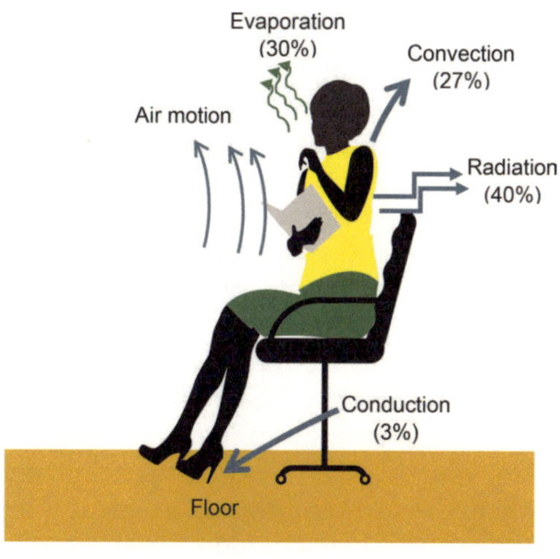

Figure 1.
Heat transfer mechanisms from a resting person (Adapted from [25]).

and radiation. Radiation is neglected in the lungs because the heat is transferred by convection, as air is continuously moving in and out during breathing. The contribution of radiation is minimal due to the small surface area of the lungs and the limited temperature difference between the lungs and the surrounding air, making it insignificant compared to convection [25]. The following equations represent these processes:

$$\dot{Q}_{\text{sensible,skin}} = \dot{Q}_{\text{conv,skin}} + \dot{Q}_{\text{rad,skin}} \tag{5}$$

$$\dot{Q}_{\text{sensible,lungs}} = \dot{Q}_{\text{conv,lungs}} \tag{6}$$

Latent heat is the energy used to evaporate water from the skin and lungs using body heat. This energy is released when the water vapour condenses on cold surfaces.

Sensible heat transfer in the lungs occurs when the inhaled air is warmed, depending on the rise in air temperature. Sensible heat loss from the skin depends on factors like the skin's temperature, the air around it, nearby surfaces, and the movement of the air. Latent heat loss, on the other hand, is affected by how moist the skin is and the humidity of the air. The total rate of heat transfer expressed in W from the body to the environment can be written as:

$$\dot{Q}_{\text{body,tot}} = \underbrace{\left(\dot{Q}_{\text{sensible,skin}} + \dot{Q}_{\text{sensible,lungs}}\right)}_{\dot{Q}_{\text{sensible,tot}}} + \underbrace{\left(\dot{Q}_{\text{latent,skin}} + \dot{Q}_{\text{latent,lungs}}\right)}_{\dot{Q}_{\text{latent,tot}}} \tag{7}$$

2.2 Convective heat transfer

Convective heat transfer involves moving heat from the skin to the air around it. The rate of heat transfer through convection from the skin to the ambient can be represented by the following equation:

$$\dot{Q}_{\text{conv}} = h_c \cdot S_{\text{unclo}} \cdot \left(\overline{T}_{\text{skin}} - T_{\text{amb}}\right) \tag{8}$$

where $\overline{T}_{\text{skin}}$ is the mean skin temperature, T_{amb} is the temperature of the ambient air that refers to the temperature of the air immediately surrounding the body (is the dry-bulb temperature of the air surrounding the occupant), S_{unclo} is the unclothed body area that represents the surface area of the body that is not covered by clothing and is exposed to the surrounding air, in m^2, and h_c is the convective heat transfer coefficient in $W/(m^2{\cdot}K)$.

The convective heat transfer coefficient h_c is defined as "the net sensible heat transfer per unit area between a surface and a moving fluid medium per unit temperature difference between the surface and the medium" according to the Standard ISO 13731:2001 [26].

When the temperature difference between the skin and the surrounding air increases, heat is lost faster through convection. This happens because a high-temperature difference makes the air move more, increasing the heat transfer.

The heat from the skin, when covered by clothing, is initially transferred to the clothing and subsequently to the surrounding environment. Clothing works as insulation and reduces both types of heat loss. Heat lost through respiration depends on breathing rate, the volume of inhaled air, and environmental conditions affecting heat loss through the skin. For a person wearing clothes, sensible heat is first transferred from the skin to the clothes and then from the clothes to the surrounding environment [25]. The rate of heat transfer through convection from the outer layer of a clothed body to the ambient can be described by the following equation:

$$\dot{Q}_{\text{conv}} = h_c \cdot S_{\text{clo}} \cdot f_{\text{clo}} \cdot \left(\overline{T}_{\text{clo}} - T_{\text{amb}}\right) \tag{9}$$

where S_{clo} is the external surface of a clothed body in m^2, f_{clo} is the clothing area factor defined as the "ratio between the surface area of the clothed body, including unclothed parts, and the surface of the nude body" according to the Standard ISO 13731:2001. The clothing area factor is calculated using the equation:

$$f_{\text{clo}} = 1 + 0.3 \cdot I_{\text{clo}} \tag{10}$$

where I_{clo} represents defined in Standard ISO 13731:2001 as "the thermal insulation from the skin surface to the outer clothing surface (considering enclosed air layers)", measured in clo units ($f_{\text{clo}} = 1$ for the surface area of an unclothed body), where 1 clo = 0.155 (m^{2}°C)/W [26].

Convective heat transfer from the skin or clothing happens when air movement disturbs the insulating layer of air surrounding the body. This layer of air acts as a natural insulator, slowing down heat loss. When the layer is thick and stable, it keeps the body warm by reducing heat transfer. However, faster airflow makes this boundary layer thinner and less stable, reducing its insulating effect and allowing more heat to be lost. Convection from the skin or clothing can be classified into three types:

- Natural convection, which occurs when the air velocity, w is less than 0.2 m/s;
- Forced convection, which takes place when the air velocity, w exceeds 1.5 m/s;
- Mixed-mode convection, which occurs when the air velocity, w is between 0.2 m/s and 1.5 m/s [7].

When air flows over the human body, both natural and forced convection contribute to heat transfer.

Natural convection flow is controlled by the Grashof number, a dimensionless parameter that expresses the relationship between buoyancy forces and the viscous forces acting on the fluid:

$$\text{Gr} = \frac{g \cdot \beta \cdot \left(T_{\text{skin}} - T_{\text{amb}}\right) \cdot l^3}{\nu^2} = \frac{g \cdot \beta \cdot \Delta T \cdot l^3}{\nu^2} \tag{11}$$

where g is the gravitational acceleration, m/s^2, β is coefficient of volume expansion in K^{-1} (for ideal gases, $\beta = T^{-1}$), l is the characteristic length of the geometry, typically associated with the body height in m, and ν is the kinematic viscosity of the air, m^2/s.

The Grashof number is the key factor in determining if the fluid flow in natural convection is laminar or turbulent. If the flow is laminar, the Grashof number is less than 10^9; otherwise, if the flow is turbulent, the Grashof number exceeds this value [25]. For a stationary, unclothed body with a temperature difference of approximately 8–10°C between the mean skin temperature and the ambient, the flow remains laminar up to a height of 1 m and becomes turbulent around 1.5 m. As a result, various flow regimes occur due to natural convection at different heights, leading to varying heat loss rates across different body parts [27].

Under natural convection and for a clothed body and an unclothed body, the convective heat transfer coefficient h_c is typically in the range of 3.1 to 5.1 W/(m^2·K) [7, 27, 28]. The general equation of the convective heat transfer coefficient for a clothed body is:

$$h_c = \alpha_n \cdot \left(\overline{T}_{clo} - T_{amb}\right)^{\gamma_n} \tag{12}$$

The general equation of the convective heat transfer coefficient for an unclothed body is:

$$h_c = \epsilon_n \cdot \left(\overline{T}_{skin} - T_{amb}\right)^{\gamma_n} \tag{13}$$

where ϵ_n and γ_n are coefficients obtained from the literature under specific conditions and analysis methods [28]. These coefficients reflect the posture and surface conditions of the human body under natural convection, depending on whether the body is clothed or unclothed, with differences between sitting and standing positions. The values presented in [7] suggest that heat transfer is influenced by both posture and contact with surfaces, highlighting the role of environmental factors and body configuration in convective heat transfer processes.

In forced convection, significant research has focused on empirically determining the convective heat transfer coefficients for the entire human body. In forced convection, the Reynolds number assumes the same importance as the Grashof number in natural convection. The Reynolds number (Re), which depends on air velocity and the characteristic length related to the body, helps categorise convective flows. Its expression is:

$$\mathrm{Re} = \frac{w \cdot l}{\nu} = \frac{w \cdot l \cdot d_{air}}{\mu} \tag{14}$$

where w is the air velocity over the body surface, measured in m/s, μ is the dynamic viscosity of the air in (Pa·s), and d_{air} is the air density in kg/m3.

Recall that the Grashof number represents the ratio of buoyancy forces to viscous forces acting on the fluid, making it important for natural convection. Meanwhile, the Reynolds number is the ratio between inertial and viscous forces on a fluid, mainly used in forced convection [29]. A key criterion for distinguishing between natural and forced convection is the Archimedes number (Ar). The Archimedes number represents the ratio between the Grashof number and the square of the Reynolds number:

$$\mathrm{Ar} = \frac{\mathrm{Gr}}{\mathrm{Re}^2} \tag{15}$$

When $\frac{\mathrm{Gr}}{\mathrm{Re}^2}$ is much greater than 1, natural convection prevails, and forced convection can be neglected. In contrast, when $\frac{\mathrm{Gr}}{\mathrm{Re}^2}$ is much less than 1, forced convection dominates, and natural convection has a minor effect. However, when $\frac{\mathrm{Gr}}{\mathrm{Re}^2}$ is near 1, both natural and forced convection significantly influence the body's heat transfer, meaning that both mechanisms must be considered when analysing how heat is dissipated from the skin [25].

Mixed-mode convection for the human body refers to the combined effects of natural and forced convection in heat transfer from the body to the ambient. It typically occurs when the air velocity around the body is within the range of 0.2 m/s to 1.5 m/s, where neither natural nor forced convection is dominant. Mixed-mode convection is particularly relevant in environments where airflow is not strong enough to be fully dominated by forced convection. However, the airflow still has a significant effect on the heat exchange between the skin and the surrounding air [30]. Several factors determine the effectiveness of mixed-mode convection for the human body. These include:

- *Air velocity* - the balance between natural and forced convection depends on air velocity. If the air velocity increases beyond 0.2 m/s, forced convection (caused by air movement) plays a more significant role alongside natural convection (caused by warm air rising), resulting in increased overall heat loss from the body. As air velocity increases, forced convection becomes more important, gradually becoming the main mechanism. Once the air velocity exceeds 1.5 m/s, forced convection becomes the dominant process, and natural convection contributes less to the total heat transfer [30].

- *Body posture and orientation* - the effectiveness of mixed-mode convection also depends on the posture and orientation of the human body. Studies by [28] have shown that standing and seated postures, as well as body angles relative to airflow, can modify boundary layer dynamics, impacting the balance between natural and forced convection. A person standing up in mild airflow loses heat more uniformly across the body than a seated person, where airflow may affect certain body parts more than others, such as the legs or arms.

- *Clothing and surface conditions* - the presence of clothing can significantly modify heat transfer in mixed-mode convection. Clothing acts as an additional barrier that influences both natural and forced convection. According to the study by [31], mixed-mode convection over clothed surfaces shows higher heat loss rates than unclothed ones, particularly when air velocities are moderate. This happens because the clothing fabric allows some air to pass through, which increases forced convection. At the same time, it also permits natural convection to occur within the space between the fabric and the skin.

- *Environmental Conditions* - Indoor environments with moderate air movement, such as those caused by fans or Heating, Ventilation, and Air Conditioning (HVAC) systems, typically exhibit mixed-mode convection. Understanding how air movement and temperature differences impact the thermal comfort of occupants is essential for designing spaces that provide a comfortable environment. Both airflow and temperature gradients are crucial in ensuring thermal comfort [27].

2.3 Radiative heat transfer

Radiative heat transfer in the human body involves releasing heat from the skin through electromagnetic radiation, primarily within the infrared spectrum. This heat transfer occurs due to the difference in temperature between the body and its environment. When the skin temperature exceeds that of nearby surfaces or surrounding air, the body releases infrared radiation, carrying heat away and contributing to thermal balance and comfort.

The rate of heat transfer through radiation from clothing to surrounding air can be calculated using:

$$\dot{Q}_{\text{rad}} = h_{\text{r}} \cdot S_{\text{clo}} \cdot f_{\text{clo}} \cdot \left(\overline{T}_{\text{clo}} - \overline{T}_{\text{r}}\right) \tag{16}$$

where h_r is the radiative heat transfer coefficient, and $\overline{T}_r$ is the mean radiation temperature.

The radiative heat transfer coefficient h_r is defined as "the net rate of heat transfer per unit area by radiation between two surfaces, per unit temperature difference between the surfaces" according to ISO 13731:2001 [26]. The radiative heat transfer coefficient in $W/(m^2{\cdot}K)$ is calculated using the following formula:

$$h_{\text{r}} = 4 \cdot \overline{\varepsilon} \cdot k_{\text{B}} \cdot \left(\frac{S_{\text{r}}}{S_{\text{DU}}}\right) \cdot \left[273.2 + \frac{\overline{T}_{\text{clo}} + \overline{T}_{\text{r}}}{2}\right]^3 \tag{17}$$

where $\overline{\varepsilon}$ is the mean emissivity of the human body's surface ranging from 0 to 1 (dimensionless), $k_B = 5.67 \cdot 10^{-8}\ W/(m^2 \cdot K)$ is the Stefan-Boltzmann constant, S_r is the effective radiation area of the human body in m^2, and S_{DU} is the DuBois body surface area, "the total surface area of a nude person", which is an estimation of the total surface area of the human body based on height and weight. The formula of the DuBois body surface area is expressed as [32]:

$$S_{\text{DU}} = 0.202 \cdot m^{0.425} \cdot H^{0.725} \tag{18}$$

where m is the body weight in kg and H is the body height in m.

The mean radiant temperature, $\overline{T}_r$ reflects the combined effect of these surfaces on radiative heat exchange with the human body. This temperature plays an important role in assessing the thermal comfort of occupants in the indoor environment. The mean radiant temperature $\overline{T}_r$ is defined as the "uniform temperature of an imaginary black enclosure in which an occupant would exchange the same amount of radiant heat as in the actual non-uniform enclosure" according to ISO 13731:2001 [26]. To calculate the mean radiation temperature, each surface surrounding the human body contributes to the radiative exchange based on the view factor, $F_{\text{human body}-i}$. The view factor is a geometric quantity that does not depend on surface properties and temperature. The mean radiant temperature $\overline{T}_r$ can be estimated using the following formula [25]:

$$\overline{T}_{\text{r}} = \sum_{i=1}^{N} F_{\text{human body}-i} \cdot T_i \tag{19}$$

where $F_{\text{human body}-i}$ is the view factor defined as the fraction of the radiation leaving the human body that strikes the surface i directly without being reflected or absorbed. The view factor depends on the size, orientation, and distance of the surrounding surfaces relative to the person. A larger, closer surface will have a higher view factor than smaller, more distant surfaces, meaning this surface contributes more to the radiative heat exchange with the human body. The view factor ranges between 0 and 1.

- $F_{\text{human body}-i} = 1$ means all radiation emitted by the human body reaches the surface i;
- $F_{\text{human body}-i} = 0$ means no radiation emitted by the human body reaches the surface i.

2.4 Combined convective and radiative heat transfer

Total sensible heat loss can be calculated by combining both convective and radiative heat losses from the outer surface of a clothed body. The equation of the total heat transfer due to both convection and radiation from the clothed body can be expressed as:

$$\dot{Q} = h_{\text{tot}} \cdot S_{\text{clo}} \cdot \left(\overline{T}_{\text{clo}} - T_{\text{op}}\right) = \frac{S_{\text{clo}} \cdot \left(\overline{T}_{\text{clo}} - T_{\text{op}}\right)}{\frac{1}{h_c + h_r}} = \frac{S_{\text{clo}} \cdot \left(\overline{T}_{\text{clo}} - T_{\text{op}}\right)}{R_{\text{t,conv+rad}}} \tag{20}$$

where h_{tot} is the combined heat transfer coefficient for convection and radiation in W/(m^2·K) and T_{op} is the operative temperature in K, which represents a weighted average between the surrounding air and radiative temperatures.

The operative temperature is derived from the mean radiant temperature $\overline{T}_r$, the temperature of the ambient air T_{amb}, and air velocity over the body surface w. The operative temperature is defined as the "uniform temperature of an imaginary black enclosure in which an occupant would exchange the same amount of heat by radiation plus convection as in the actual non-uniform environment", according to the Standard ISO 13731:2001 [26]. The operative temperature is calculated as:

$$T_{\text{op}} = \frac{h_c \cdot T_{\text{amb}} + h_r \cdot \overline{T}_{\text{surr}}}{h_c + h_r} \tag{21}$$

The mean radiant temperature significantly influences the operative temperature, particularly in moderate thermal environments [33–35]. In moderate environments, conditions are comfortable and stable, with temperatures, humidity, and air movement at levels that feel natural and pleasant. Occupants in these conditions usually do not feel too hot or cold, and the temperature difference is small [36]. When the temperature difference $\overline{T}_{\text{surr}}$-$T_{\text{amb}}$ is less than 4°C, and the air velocity w is less than 0.2 m/s, the operative temperature is computed as [34]:

$$T_{\text{op}} = 0.5 \cdot \left(T_{\text{amb}} + \overline{T}_{\text{surr}}\right) \tag{22}$$

This equation is particularly used when $h_c = h_r$. Under these conditions, Eq. (22) gives a good estimate of the temperature that occupants feel in moderate environments.

Heat transfer through clothing involves the movement of the heat from the skin to the outer surface of the clothing. The rate of heat transfer through clothing or total sensible heat loss through clothing can be expressed as [25]:

$$\dot{Q} = \frac{S_{\text{clo}} \cdot \left(\overline{T}_{\text{skin}} - \overline{T}_{\text{clo}}\right)}{R_{\text{t,clo}}} \tag{23}$$

where R_{clo} is the thermal resistance of clothing in clo, where 1 clo $= 0.155\ \frac{m^2 \cdot ^\circ C}{W}$.

The total sensible heat loss for the combined heat transfer phenomena can be expressed as:

$$\dot{Q}_{sensible,skin} = \frac{S_{clo} \cdot \left(\overline{T}_{skin} - \overline{T}_{clo}\right)}{R_{t,tot}} = \frac{S_{clo} \cdot \left(\overline{T}_{skin} - T_{op}\right)}{R_{t,clo} + R_{t,conv+rad}} \tag{24}$$

where $R_{t,tot}$ is the total thermal resistance that combines the effects of conduction through the clothing and convection and radiation from the outer surface of the clothing to the surroundings of the human body, in $\frac{m^2 \cdot K}{W}$.

Thermal resistance of clothing depends on the type and thickness of clothing. **Table 2** provides the values of typical insulation for different types of clothing in clo units, along with their thermal resistance in $\frac{m^2 \cdot K}{W}$.

The sweat on the body's surface undergoes a phase-change process. Latent heat loss occurs as a result of sweat evaporation and the movement of water through the skin's external surface. Furthermore, the evaporation process achieves its highest efficiency when the skin is fully saturated with moisture [25]. The heat transfer rate (or latent/ evaporative heat losses) depends on the rate of sweat evaporation on the body surface and the enthalpy of sweat vaporisation and indicates the level of moisture present on the skin:

$$\dot{Q}_{latent,skin} = \dot{m}_{sweat\ evap} \cdot i \tag{25}$$

where $\dot{m}_{sweat\ evap}$ is the rate of sweat evaporation on the body surface (skin), in kg/s and i is the enthalpy of sweat vaporisation at the specified temperature or pressure in J/kg.

During respiration, the air inhaled enters the lungs under ambient conditions, while the exhaled air is saturated with moisture at a temperature close to the body's core temperature. As a result, the body loses heat through both sensible heat transfer by convection and latent heat transfer due to evaporation from the lungs. The rate of heat transfer or sensible heat loss through convection is described as follows:

$$\dot{Q}_{sensible,lungs} = \dot{m}_{air,lungs} \cdot c_{p,air} \cdot \left(T_{exh} - T_{amb}\right) \tag{26}$$

where $\dot{m}_{air,lungs}$ is the mass flow rate of air through the lungs in kg/s; $c_{p,air}$ is the specific heat capacity of air in J/(kg·K), and T_{exh} is the temperature of the exhaled air, in K.

The rate of heat transfer or latent heat loss through evaporation is described as follows [25]:

Clothing type	Clothing insulation I_{cl}, clo	Thermal resistance R_{clo}, $\frac{m^2 \cdot K}{W}$
Trousers, T-shirt, sweater, long-sleeve shirt	1.0	0.1550
Summer clothing	0.5	0.0775
Winter clothing	0.9	0.1395

Table 2.
Clothing insulation and Thermal resistance value for different types of clothing [25].

$$\dot{Q}_{\text{latent,lungs}} = \dot{m}_{\text{air,lungs}} \cdot \lambda \cdot (u_{\text{exh}} - u_{\text{amb}}) \tag{27}$$

where λ is the latent heat of vaporisation of water in J/kg, which is the amount of heat required to convert one unit of mass of a liquid into vapour (or vapour into liquid) at constant temperature and pressure without any change in the substance's temperature; *u* is the humidity ratio, also known as specific humidity, which represents the ratio of water vapour mass compared to the mass of dry air in an air mixture. It applies to the exhaled air (u_{exh}) and ambient air (u_{amb}) and is dimensionless.

For exhaled air, the moisture comes from water vapour released by the respiratory tract, while for ambient air, it is influenced by the surrounding environmental humidity [37]. The humidity ratio is expressed as:

$$w = \frac{m_{\text{moisture}}}{m_{\text{dry air}}} \tag{28}$$

where m_{moisture} is the mass of water vapour present in the air (the exhaled air in the case of w_{exh}, and the ambient air in the case of w_{amb}) in kg, and $m_{\text{dry air}}$ is the mass of dry air in kg.

In conclusion, convective and radiative heat transfer are important for understanding how the body loses heat through clothing. These processes depend on air temperature, clothing properties, and the surrounding environment. By considering these factors together, it is possible to understand better how to manage heat losses and improve comfort in everyday situations.

3. Thermal interactions between human skin and wearable thermoelectric systems

Knowledge of heat transfer principles in the human body is useful for assessing the thermal interactions between human skin and wearable thermoelectric systems. For this reason, it is important to analyse how these systems interact with the body's natural heat transfer mechanisms to regulate or harvest heat effectively.

Wearable thermoelectric generators (w-TEGs) operate by harvesting energy from the temperature difference between the human body and the surrounding environment. This section explains the heat transfer processes in human skin (Section 3.1) and their relevance to the operation of w-TEGs, as well as the analytical equations governing heat transfer in w-TEGs (Section 3.2).

3.1 Analytical heat transfer in human skin

In developing analytical equations for heat transfer related to the interaction between human skin and w-TEGs, the skin is modelled as a quasi-homogeneous structure with four distinct layers (the inner layer is the muscle, followed by the fat, dermis, and the external layer is the epidermis). Each layer has a different thickness, δ and unique thermal properties, including thermal conductivity. **Figure 2** illustrates the components of the w-TEG system (flexible substrate, the w-TEG embedded in the filling material, the encapsulating layer, and the heat sink spreader) and the layers of the human body. The heat flux, $\dot{Q}_{\text{h}}$ moves from the skin surface ($T_{\text{skin}} \approx 33°\text{C}$) towards the w-TEG, as indicated by the red arrow. The blue arrow represents the heat flux $\dot{Q}_{\text{c}}$

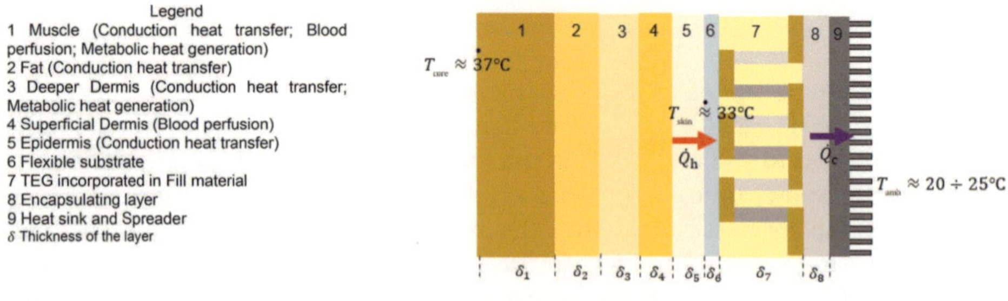

Figure 2.
Schematic representation of the w-TEG device showing the layered structure from the human body to the ambient environment.

dissipated through the encapsulating layer and the heat sink to the ambient air ($T_{\text{amb}} \approx 20 \div 25^{\circ}\text{C}$).

The following equation governs the steady-state heat conduction through each layer [38]:

$$\frac{d^2T_j}{dx^2} = -\left(\dot{q}_{\text{met}} + \dot{q}_{\text{perfusion}}\right) \cdot \frac{1}{k_j} \tag{29}$$

where T_j is the temperature of the skin layer j (with $j = 1 \div 5$), $\dot{q}_{\text{met}}$ is the volumetric heat generation rate by metabolism in W/m^3, $\dot{q}_{\text{perfusion}}$ is the volumetric heat generation rate due to the blood perfusion in W/m^3, and k_j is the thermal conductivity of the layer j (with $j = 1 \div 5$) in W/(m·K).

The heat transfer associated with blood flow in tissues is expressed as:

$$\dot{q}_{\text{perfusion}} = \left(T_{\text{core}} - T_j\right) \cdot d_{\text{fluid}} \cdot c_{\text{p fluid}} \cdot \gamma \tag{30}$$

where T_{core} is the core temperature in K, d_{fluid} is the blood density in kg/m^3, c_{pfluid} is the specific capacity of blood at constant pressure in J/(kg·K), $\gamma = \dot{V}_{\text{fluid}} \cdot \left(V_{\text{tissue}} \cdot \Delta t\right)^{-1}$ in s^{-1} is the blood perfusion coefficient, representing the blood flow per unit tissue volume over time, $\dot{V}_{\text{fluid}}$ is the blood flow rate in a particular tissue in m^3/s, V_{tissue} is the volume of the tissue where blood flow occurs in m^3 and Δt is the time interval during which the blood flow is measured in s.

The perfusion coefficient γ is important for understanding localised blood circulation and is used to regulate the body temperature. When the blood flow to the skin increases, the process is called vasodilation, which enhances heat release from the body. A reduction in blood flow is called vasoconstriction, minimising heat loss. Both mechanisms are critical for keeping the thermal balance and ensuring proper physiological function of the body [39].

In the thermal assessment of the fat and epidermis layer, the impact of blood perfusion is neglected ($\dot{q}_{\text{perfusion}} = 0$). For this reason, the steady-state heat conduction within the tissue layers becomes:

$$\frac{d^2T_j}{dx^2} = -\frac{\dot{q}_{\text{met}}}{k_j} \tag{31}$$

For the fat and epidermis layers (j = 2;5), the solution of the temperature field takes the form of a quadratic function:

$$T_j\ (x) = -\frac{\dot{q}_{\text{met}}}{k_j} \cdot x^2 + \text{A}_1 \cdot x + \text{A}_2 \tag{32}$$

where A_1 and A_2 are the integration constants which are obtained from the boundary conditions.

For the dermis layer (j = 3;4), the solution of the temperature fields is more complex as it considers the effects of blood perfusion, which has an important role in heat dissipation. The expression of the temperature field is an exponential function:

$$T_j\ (x) = T_{\text{core}} - \left(\rho_{\text{fluid}} \cdot c_{\text{p fluid}} \cdot \gamma\right)^{-1} \cdot \dot{q}_{\text{met}} \cdot e^{r \cdot x} + \text{A}_3 \cdot e^{r \cdot x} + \text{A}_4 \cdot e^{-r \cdot x} \tag{33}$$

where A_3, A_4 are the integration constants which are obtained from the boundary conditions and r is a parameter that depends on the tissue properties and blood perfusion, defined as:

$$r = \sqrt{\frac{\rho_{\text{fluid}} \cdot c_{\text{p fluid}} \cdot \gamma}{k_j}} \tag{34}$$

The integration constants A_1, … , A_4 for the fat, epidermis and dermis layers are determined using the boundary conditions for the respective temperature distributions in these homogeneous tissues [38, 40]. These constants are directly influenced by the thermal properties of the layers and the heat generation rates. Advanced modelling techniques improve the integration of thermoelectric systems in wearable applications. Wearable TEGs can significantly enhance energy harvesting efficiency through optimised design and thermal management [40]. Section 3.2 introduces the main equations and principles for steady-state heat transfer in w-TEGs, focusing on thermal conductivities, temperature profiles, and heat flux calculations at different device interfaces.

3.2 Analytical heat transfer equations for wearable thermoelectric generators

To assess the thermal behaviour of w-TEG, it is assumed that the thermoelectric legs (P-type and N-type) have the same material properties and dimensions. This assumption allows for calculating the simplified steady-state equations to obtain the temperature distribution and the heat fluxes absorbed and released through the device. Under this assumption, in the theoretical analysis, only the N-type thermoelectric legs are written in the heat transfer equations [38, 40].

The steady-state temperature distribution in the thermoelectric legs is described by the following equation (as shown in **Figure 3**):

$$\frac{\text{d}^2 T_{\text{N}}}{\text{d}x^2} + \frac{I^2 \cdot R_{t\ N}}{K_N \cdot \delta^2_{\text{TEG+fmat}}} = 0 \tag{35}$$

where I is the electric current in the w-TEG in A, R_N is the electric resistance of the N-type thermoelectric leg in Ω, K_N is the thermal conductance of the N-type thermoelectric leg in $W \cdot K^{-1}$, $R_{t\ N}$ is the thermal resistance of the N-type thermoelectric leg in $K \cdot W^{-1}$ and $\delta_{\text{TEG+fmat}}$ is the thickness of the combined TEG and fill material in m.

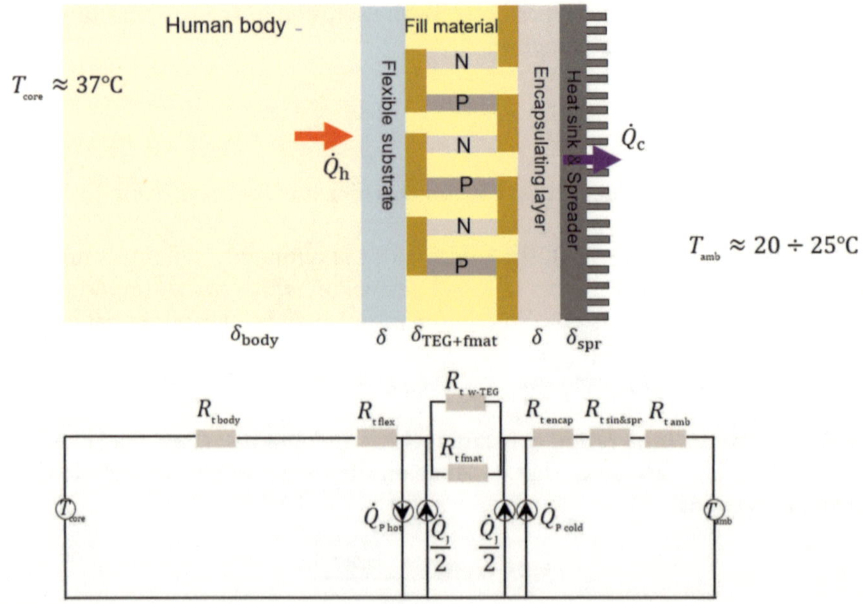

Figure 3.
Illustration of the thermoelectric energy harvesting utilising body heat. The figure also depicts the thermal equivalent circuit used in the theoretical analysis, where $\dot{Q}_{h\ Peltier}$ and $\dot{Q}_{c\ Peltier}$ represent the heat contributions due to the Peltier effect on the hot and cold sides, respectively.

Unlike thermoelectric legs, Joule heating in metallic electrodes is negligible and, therefore, not included here [38]. The electric current does not pass through the flexible substrate, and the electric current, I is set to zero ($I = 0$). Under this condition, Eq. (35) simplifies to:

$$\frac{d^2 T_N}{dx^2} = 0 \tag{36}$$

The simplified representation of the thermal circuit depicted in **Figure 3** is described in [7, 38, 40]. The optimal thermal resistance for the thermal circuit can be expressed as:

$$R_{t\ opt} = \frac{1}{2 \cdot R_{t\ fmat}{}^{-1} + \left(R_{t\ body} + R_{t\ amb}\right)^{-1}} \tag{37}$$

where $R_{t\,fmat}$ is the parasitic thermal resistance introduced by the fill material, which is connected in parallel with the overall thermal resistance of the w-TEG containing n thermoelements of N-type and P-type, $R_{t\ body}$ is the thermal resistance of the body, and $R_{t\ amb}$ is the thermal resistance of the ambient air in K·W^{-1}.

The overall thermal resistance of the w-TEG system is:

$$R_{t\ w-\mathrm{TEG}} = \frac{1}{FF} \cdot \frac{l}{k \cdot S_{w-\mathrm{TEG}}} \tag{38}$$

where FF is the fill factor, defined as the ratio of the total area of the thermoelectric legs ($S_N + S_P$) to the total surface area of the w-TEG system $S_{w-\mathrm{TEG}}$, calculated as

$FF = \frac{S_N + S_P}{S_{w-\text{TEG}}}$; k is the total thermal conductivity of the thermoelectric legs ($k = k_N + k_P$) in W/(m·K), $S_{\text{w}-\text{TEG}}$ is the cross-sectional area of the thermoelectric legs in m^2, and l is the length of the thermoelectric legs in m.

The heat flux absorbed at the interface between skin and w-TEG, considering the flexible substrate, is expressed as:

$$\dot{Q}_h = \frac{T_{\text{skin}} - T_{\text{flex}}}{R_{t\ \text{eq}\ 1}} \tag{39}$$

where $R_{\text{t eq 1}} = R_{\text{t body}} + R_{\text{t flex sub}}$ is the equivalent thermal resistance, $R_{\text{t flex sub}}$ is the thermal resistance of the flexible substrate, and T_{flex} is the temperature at the flexible substrate in K.

The heat flux released through the encapsulating layer and the heat spreader at the heat sink is defined as:

$$\dot{Q}_c = \frac{T_{\text{encapt}} - T_{\text{amb}}}{R_{\text{eq}\ 2}} \tag{40}$$

where $R_{\text{eq 2}} = R_{\text{encap}} + R_{s\text{in\&spr}} + R_{t\ \text{amb}}$ is the equivalent thermal resistance, R_{encap} is the thermal resistance of the encapsulating layer, $R_{s\text{in\&spr}}$ is the thermal resistance of the heat sink&spreader, T_{encapt} is the temperature of the encapsulating layer in K.

According to the principles of energy conservation under steady-state conditions, the value of the heat flux across different layers (fat, dermis, epidermis, flexible substrate, thermoelectric legs, encapsulating layer, and heat sink) does not change. This condition is expressed as follows:

$$\dot{Q}_{\text{layer}}(\delta_i) = \dot{Q}_{\text{next layer}}(\delta_i) \tag{41}$$

where $\dot{Q}_{\text{layer}}$ is the rate of heat transfer across a specific layer in W, $\dot{Q}_{\text{next layer}}$ is the rate of heat transfer across the adjacent layer in W, and δ_i is the thickness of the i-th layer in m. Specifically:

$$\dot{Q}_{\text{skin}-w-\text{TEG}}(\delta_{\text{skin}}) = \dot{Q}_{\text{flex}}(\delta) \tag{42}$$

$$\dot{Q}_{\text{layer}}\left(\delta_{\text{encap}}\right) = \dot{Q}_{w-\text{TEG}+f\text{mat}}(\delta_{w-\text{TEG+fmat}}) \tag{43}$$

$$\dot{Q}_{w-\text{TEG+fmat}}(\delta_{w-\text{TEG+fmat}}) = \dot{Q}_{\text{spr}}\left(\delta_{\text{spr}}\right) \tag{44}$$

These expressions are useful for obtaining the temperatures at the hot and cold sides of the w-TEG, as well as intermediate temperatures within the w-TEG layers. A detailed discussion of the temperature distributions and heat flux expressions can be found in [38, 40].

4. Performance of wearable thermoelectric generators

The performance of w-TEGs relies on harnessing the temperature gradient between a heat source (i.e., the human body) and a heat sink (i.e., the ambient environment). A theoretical assessment involves evaluating the power generated based on the principles of energy conservation. The output power of the w-TEG is expressed as the difference between the net heat fluxes of the hot and cold sides:

$$P_{\text{out}} = \dot{Q}_h - \dot{Q}_c \tag{45}$$

By considering the contributions of the thermoelectric effects, the expression of the output power is:

$$P_{\text{out}} = n \cdot (\alpha_{\text{P}} - \alpha_{\text{N}}) \cdot (T_h - T_{\text{c}}) \cdot I - \frac{R \cdot I^2}{2} = \underbrace{n \cdot \alpha \cdot \Delta T}_{V_{\text{oc}}} \cdot I - \frac{R \cdot I^2}{2} \tag{46}$$

where R is the electric resistance of the thermoelectric module, $\Delta T = T_h - T_{\text{c}}$ is the temperature difference between the hot and cold sides of the w-TEG in K, T_h is the temperature of the hot side of the w-TEG in K, T_{c} is the temperature of the cold side of the w-TEG in K, V_{oc} is the open circuit voltage generated by the Seebeck effect in V, α is the Seebeck coefficient of the thermoelements (α_{P} of P-type thermoelement and α_{N} of the N-type thermoelement) in V·K^{-1}, and n is the number of thermoelectric elements.

When a load resistance, R_L is connected to the system, the output power is:

$$P_{\text{out}} = R_L \cdot I^2 \tag{47}$$

By equating Eq. (46) with Eq. (47), it is obtained the expression of the electric current:

$$I = \frac{n \cdot \alpha \cdot \Delta T}{R + R_L} \tag{48}$$

where R is the internal resistance in Ω.

Eq. (47) can be also written as:

$$P_{\text{out}} = R_L \cdot \frac{V_{\text{oc}}{}^2}{(R + R_L)^2} \tag{49}$$

Moreover, when $R = R_{\text{L}}$, the output power has the maximum value as:

$$P_{\text{out}} = \frac{V_{\text{oc}}{}^2}{4 \cdot R_L} \tag{50}$$

If the temperatures of the hot and cold sides are substituted in Eq. (47), a cubic equation for the electric current is obtained [38, 40, 41]:

$$\text{B}_1 \cdot I^3 + \text{B}_2 \cdot I^2 + \text{B}_3 \cdot I + \text{B}_4 = 0 \tag{51}$$

where B_1, B_2, B_3, and B_4 are known coefficients, as presented in [7]. The solution of this cubic equation provides the optimal current for a given load and temperature difference, which is then used to calculate the output power.

Another parameter to evaluate the performance of the w-TEG is the maximum power density:

$$P_{d\ \max} = \frac{\Delta T^2}{4 \cdot L} \cdot \frac{\alpha^2 \cdot FF}{\rho} \tag{52}$$

where ρ is the electrical resistivity in $\Omega \cdot m$.

Pang et al. [42] analysed the output power density of the flexible TEG by varying the fill factor *FF* and the thermoelectric element lengths. They found that increasing the fill factor initially improves power density by lowering electrical resistance and enhancing thermal conductance. However, after obtaining the optimal fill factor, the temperature difference across the device decreases more significantly than the reduction in resistance, leading to a reduction in power density. The key parameters, such as the load resistance, the fill factor, the thermophysical properties of the filler material, and the thermal behaviour of the human skin, influence the performance of w-TEGs.

The performance of w-TEGs is also influenced by integrating flexible materials that can adapt to dynamic human motion and improve personal comfort. Section 5 discusses the importance of flexibility in w-TEGs, highlighting their role in ensuring adaptability to body movements, efficient heat harvesting, and user comfort.

5. Flexibility of wearable thermoelectric generators

The flexibility of w-TEGs is essential to harvest the maximum amount of wasted heat from the body, as these devices must conform to the shape of the body [43]. Flexibility is also essential to w-TEGs for maintaining the wearer's comfort and maximising the efficiency of heat harvesting from the human body. Since w-TEGs are intended to be worn directly on or near the skin, they must be able to adapt to the natural movements of the human body without causing discomfort. A rigid w-TEG would not only be uncomfortable but also reduce the contact area with the skin, limiting its ability to absorb body heat effectively. Traditional TEGs cannot be used effectively on curved surfaces due to their rigid structural components.

The following methods are used to achieve flexibility of w-TEGs:

- Designing TEGs with flexible substrates and electrodes [44] ensures adaptability to the human body's curved surfaces [45];
- Use of flexible thermoelectric materials as thermoelectric legs;
- Incorporating w-TEGs into textile clothing to enhance adaptability [43].

5.1 Designing TEGs with flexible substrates and electrodes

Wearable TEGs are designed to be placed on different parts of the human body (e.g., wrist, forearm, legs, forehead, chest, etc). Wearable TEGs are typically designed with rigid thermoelectric legs on a flexible substrate [46]. The flexible substrate is the material on which the thermoelectric legs are mounted, providing structural support and allowing the w-TEG to conform to the curved surfaces of the body. The choice of flexible substrate material is important because a material with low thermal conductivity $k_{\text{flex sub}}$ increases the thermal resistance of the flexible substrate $R_{t\ \text{flex sub}}$ between the skin and the w-TEG. A higher $R_{t\ \text{flex sub}}$ reduces the heat flux reaching the w-TEG, $\dot{Q}_{\text{flex}}$, which decreases the temperature difference between the TEG sides ΔT, ultimately lowering the voltage and output power of the w-TEG. Sakamoto et al. [47] introduced an experimental method to assess the thermal resistance of the interface layer (TIM) placed when TEG is placed on a flat surface. TIM materials (i.e., BNbased ceramic paste and TSU700-H) effectively reduce thermal contact resistance, enhancing TEG performance at medium temperatures. When the w-TEG is applied to the curved skin area, the flexible substrate adjusts its shape by bending and deforming to conform to the skin's contours. Flexibility ensures better contact between the w-TEG and the skin, enabling more efficient heat transfer despite the non-uniform surface (**Figure 4**).

The shape and curvature of the skin can significantly influence the performance of the w-TEG. For instance, the w-TEG must bend to match the skin's curvature, which results in non-uniform temperature distribution across the flexible substrate. The material properties of the substrate, such as thickness and thermal conductivity, play an important role in the w-TEG's efficiency. To enhance performance, heat loss in the flexible substrate must be minimised. Wang et al. [48] developed a numerical model to study how the thickness δ and thermal conductivity of the flexible substrate affect w-TEG performance on curved surfaces. Their study showed that using a thinner flexible substrate with higher thermal conductivity improves heat transfer and voltage generation of w-TEG for human body harvesting. Eom et al. [49] developed a flexible TEG with a bracelet-like modular design that combines rigid inorganic materials with polymer links and a copper heat sink. This design allows the TEG to conform to surfaces like the wrist while improving heat transfer and efficiency. In their experiments, the flexible TEG generates about 80 μW powered by human body heat during light activity and usable output power even at rest. However, high thermal contact resistance from thick copper electrodes limited w-TEG performance.

Electrodes play a key role in ensuring both electrical conductivity and flexibility in w-TEGs. Kunag et al. [50] proposed an innovative heteromorphic electrode design integrated with a heat sink to enhance heat dissipation at the cold side of w-TEG. The design significantly improved the maximum output power, achieving a 44.1% increase compared to conventional electrodes. Their work demonstrated the potential of flexible TEGs with advanced electrodes for efficient energy harvesting, obtaining a power density of 21.3 μW/cm^2 under natural convection and 116.1 μW/cm^2 with forced convection, making them practical for wearable applications. Thinner, flexible electrodes were identified as a solution to improve efficiency, highlighting the need for improved flexible materials and design. Selecting electrode materials

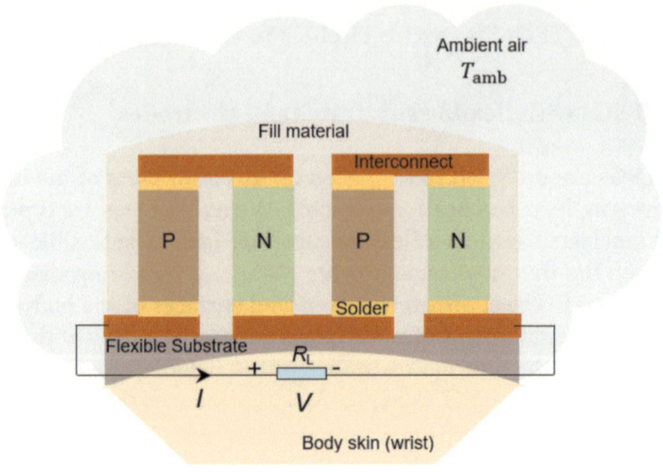

Figure 4.
Illustration of the TEG structure positioned on a curved skin surface (Adapted from [48]).

(e.g., conductive polymers and metal-based composite) with high electrical conductivity and mechanical flexibility reduces resistive losses while ensuring user comfort [51].

5.2 Flexible thermoelectric materials as thermoelectric legs

Flexible thermoelectric materials, such as conductive polymers or carbon-based composites, are designed to maintain good thermoelectric performance while being flexible enough to move with the body. The substrates, which are the base layers that hold the thermoelectric elements, must also be flexible to ensure the entire device can adapt to the wearer's movements.

Wearable TEGs can be classified into two types: w-TEGs made from organic materials that are naturally flexible, and w-TEGs made from inorganic bulk materials with flexible design [52].

Flexible organic thermoelectric materials [53] are more comfortable and softer, but their lower performance limits their ability to provide stable and long-lasting sensor operation. Flexible organic materials use thin-film technologies like poly(3,4-ethylenedioxythiophene):poly(styrenesulfonate), known as PEDOT: PSS [54, 55]. PEDOT:PSS is the most popular conducting polymer used in w-TEGs due to its ability to maintain high Seebeck coefficients, facilitating efficient charge transport. PEDOT: PSS can be exposed to air and moisture without significant degradation [53]. Other conducting polymers include polyaniline (PANi), polypyrrole (PPY), polythiophene (PTH), polyphenylenevinylene (PPV), and polycarbazoles (PC) [8]. Compared to inorganic thermoelectric materials, these polymers have lower thermal conductivity, which is advantageous for achieving a high dimensionless figure of merit $Z\overline{T}$. Additionally, conducting polymers are lightweight, cost-effective, and easy to produce and fabricate into various forms, such as thin films, fibres, coatings, or 3D shapes [8]. Most conducting polymers and their thermoelectric composites are P-type thermoelectric materials because conventional N-type conducting polymers lack stability when air exposure. This limitation significantly impacts the advancement of wearable TEGs. The N-type conducting polymers are unstable when exposed to air, which limits their use in w-TEGs. Improving the electrical conductivity of these materials is key to increasing $Z\overline{T}$. While some progress has been made, with $Z\overline{T}$ values, challenges like air instability, difficult manufacturing processes, and low conductivity still exist. Therefore, creating N-type conducting polymers that are stable in air, easy to process in solution and exhibit high thermoelectric performance, remains a significant challenge [8].

Inorganic bulk materials have poor flexibility compared to organic thermoelectric materials. Flexible substrates overcome the limited flexibility and comfort of inorganic bulk materials. PDMS (polydimethylsiloxane) is a flexible, biocompatible silicone-based polymer used in w-TEGs as a flexible substrate or encapsulation material [56]. PDMS has low thermal conductivity and is used in designs where thermal insulation is needed to maintain a temperature gradient across the thermoelectric elements. PDMS is often used to encapsulate thermoelectric elements, protecting them from environmental factors like moisture, dust, and mechanical damage. PDMS is preferred due to its biocompatibility and non-toxic nature, making it safe for prolonged skin contact. Furthermore, PDMS is resistant to many chemicals, solvents, and environmental factors, ensuring satisfactory durability also in harsh conditions [48].

Moreover, thermoelectric materials can be divided into 1D, 2D and 3D types:

- 1D materials, like nanowires and fibres, have been widely researched but are not fully developed for practical use, especially in the medical field, due to unresolved challenges such as catalyst selection,

- 2D materials include thermoelectric thin films. Organic 2D materials, like conducting polymers, are good candidates for flexible applications. Inorganic 2D materials have some flexibility, making them promising for w-TEGs.

- 3D materials are primarily bulk semiconductors, such as bismuth telluride (Bi_2Te_3), which remains the best thermoelectric material at room temperature due to its high performance [52].

5.3 Incorporating w-TEGs into textile clothing

Integrating TEGs into textile clothing can help harvest body heat over a larger area, converting it into electrical energy. Adding TEGs to fabrics does not significantly reduce their power output. For example, a shirt with hidden TEGs placed between fabric layers can generate between 0.5 mW and 5 mW, depending on the ambient temperature (from 15 to 27°C). This demonstrates the potential for TEGs to provide an efficient energy source for w-TEGs without compromising the functionality or comfort of the clothing [57]. Du et al. [5] developed a fabric-based TEG by connecting multiple PEDOT: PSS-coated fabric pieces in series. This fabric TEG has a 2D flat structure that harvests heat along the in-plane direction. In real-life wearable applications, there is a temperature difference between the human body and the surrounding environment, which causes heat flow to move perpendicularly to the skin. A 2D flat TEG does not work well because it is parallel to the skin, preventing a temperature difference from forming along its length. Wu and Hu [58] presented a novel design for a w-TEG based on a 3D fabric structure as the substrate. This design enables the generation of a temperature difference across the thickness of the fabric, which improves its applicability for body heat harvesting in wearable systems. Chen et al. [59] highlighted that textiles, due to their structures of fibres, yarns and fabrics, are flexible, allow air to pass through, and are easy to shape. These characteristics make them perfect for adding w-TEGs without affecting comfort. Khan et al. [49], explained the advantages of fibre-shaped w-TEGs (i.e., *flexibility, durability, and ability to withstand bending and stretching*), making them ideal for integration into a textile material. To improve performance, Wu et al. [60] have applied techniques such as Ag nanowire coating, which enhance infrared reflectivity and increase voltage output by 35%. Furthermore, encapsulation helps to retain heat on the hot side, ensuring stable operation during different activities, including sports and underwater use. Other advancements include π-type thermoelectric modules made with alternately doped carbon nanotube fibres, achieving a peak power density, $P_d = 70\ \text{mW/m}^2$ at the temperature difference $\Delta T = 44$ K. Khan et al. [49] also noted recent innovations, such as the large-scale production of hierarchically segmented thermoelectric strings, which offer excellent stretchability (up to 100%), mechanical stability, and washability. Despite their benefits—lightweight design, flexibility, and scalability—fibre-based TEGs still face challenges, including low power density, complex fabrication processes, and cost limitations.

6. Conclusions

Wearable TEGs represent a promising technology for sustainable energy harvesting, especially for wearable electronic devices. By utilising the natural heat generated by the human body, w-TEGs can provide a reliable power source without depending on external energy supplies. This chapter has provided an overview of the basic principles and developments in this field. The main aspects include:

- The analysis of the mechanisms of conduction, convection, radiation, and evaporation, highlighting their role in maintaining thermoregulation and relevance to w-TEG functionality.

- A detailed discussion on the interaction of w-TEGs with the human body's natural heat transfer processes and the equations governing heat transfer in skin and w-TEG devices.

- The importance of using flexible materials and innovative designs to ensure comfort, adaptability, and efficiency in wearable applications.

By addressing aspects such as material optimisation, device flexibility, and energy management, w-TEGs can potentially become a widely used solution for powering wearable electronics.

Nomenclature

A	integration constant
Ar	Archimedes number (dimensionless)
c_p	specific heat capacity, J/(kg·K)
d	density, kg/m^3
$F_{\text{human body}}$	view factor (dimensionless)
FF	fill factor (dimensionless)
f_{clo}	clothing area factor, (dimensionless)
Gr	Grashof number (dimensionless)
g	gravitational acceleration, m/s^2
H	height, m
HVAC	heating, ventilation, and air conditioning
λ	latent heat, J/kg
h	heat transfer coefficient, W/(m^2·K)
i	enthalpy, J/kg
I	electric current, A
I_{clo}	thermal insulation, clo
K	thermal conductance, W·K^{-1}
k	thermal conductivity, W/(m·K)
k_{B}	Stefan-Boltzmann constant, $k_{\text{B}} = 5.67 \cdot 10^{-8}$ W/(m^2 · K)
l	length, m
m	mass, kg
$\dot{m}$	mass flow rate, kg/s

n	number of thermoelectric elements
P_d	power density, W/m^2
PDMS	polydimethylsiloxane
PEDOT:PSS	poly(3,4-ethylenedioxythiophene):poly(styrenesulfonate)
P_{out}	output power of the w-TEG, W
$\dot{Q}$	rate of heat transfer, W
$\dot{q}$	volumetric heat generation rate, W/m^3
r	parameter
R	electric resistance, Ω
R_t	thermal resistance, $K \cdot W^{-1}$
R_{clo}	thermal resistance of clothing, clo
Re	Reynolds number
S	surface area, m^2
t	time, s
$\overline{T}$	mean absolute temperature, K
T	absolute temperature, K
w-TEG	wearable thermoelectric generator
u	specific humidity (dimensionless)
V_{oc}	open circuit voltage generated by the Seebeck effect, V
$\dot{V}$	volumetric flow rate, m^3/s
V	volume, m^3
w	velocity, m/s
Z	figure of merit, K^{-1}

Greek symbols

α	Seebeck coefficient, $V \cdot K^{-1}$
β	coefficient of volume expansion, K^{-1}
γ	coefficient
δ	thickness, m
$\overline{\varepsilon}$	mean emissivity (dimensionless)
ϵ	coefficient
η	efficiency
μ	dynamic viscosity of the air, Pa·s
ν	kinematic viscosity of the air, m^2/s
ρ	electrical resistivity, Ω·m
σ	electrical conductivity, S/m

Author details

Diana Enescu[1,2]

1 Valahia University of Targoviste, Targoviste, Romania

2 INRIM, Torino, Italy

*Address all correspondence to: diana.enescu@valahia.ro; d.enescu@inrim.it

References

[1] McArdle WD, Katch FI, Katch VL, editors. Exercise Physiology: Nutrition, Energy, and Human Performance. 9th ed. New York, NY: Lippincott Williams and Wilkins; 2023. p. 1176

[2] Riemer R, Shapiro A. Biomechanical energy harvesting from human motion: Theory, state of the art, design guidelines, and future directions. Journal of NeuroEngineering and Rehabilitation. 2011;**8**(1):22. DOI: 10.1186/1743-0003-8-22

[3] Duan J, Liu Z, Wang X, Li TJ, Wang Y. Recent advances in skin waste heat energy harvesting wearable flexible thermo-electric and moist-electric devices. Renewable and Sustainable Energy Reviews. 2024;**202**:114719. DOI: 10.1016/j.rser.2024.114719

[4] Winter DA. Biomechanics and Motor Control of Human Movement. 4th ed. Hoboken, New Jersey, NJ: John Wiley and Sons; 2009. p. 370

[5] Du Y, Cai KF, Chen S, Wang H, Shen SZ, Donelson R, et al. Thermoelectric fabrics: Toward power-generating clothing. Scientific Reports. 2015;**5**:6144. DOI: 10.1038/srep06411

[6] Van Wylen JG, Sonntag ER. Fundamentals of Classical Thermodynamics. 3rd ed. Singapore: John Wiley and Sons; 1985. p. 772

[7] Enescu D. Heat transfer mechanisms and contributions of wearable Thermoelectrics to personal thermal management. Energies. 2024;**17**(2):285. DOI: 10.3390/en17020285

[8] Du Y, Xu J, Paul B, Eklund P. Flexible thermoelectric materials and devices. Applied Materials Today. 2018;**12**: 366-388. DOI: 10.1016/j.apmt.2018.07.004

[9] Rowe DM. Review, thermoelectric waste heat recovery as a renewable energy source. International Journal of Innovations in Energy Systems and Power. 2006;**1**(1):13-23

[10] Joshi A, Psikuta A, Bueno M-A, Annaheim S, Rossi RM. Analytical clothing model for sensible heat transfer considering spatial heterogeneity. International Journal of Thermal Sciences. 2019;**145**:135-140. DOI: 10.1016/j.ijthermalsci.2019.05.005

[11] Xu J, Psikuta A, Li J, Annaheim S, Rossi RM. Evaluation of the convective heat transfer coefficient of human body and its effect on the human thermoregulation predictions. Building and Environment. 2021;**196**:120-126. DOI: 10.1016/j.buildenv.2021.107778

[12] Havenith G, Bröde P, den Hartog E, Kuklane K, Holmer I, Rossi RM, et al. Evaporative cooling: Effective latent heat of evaporation in relation to evaporation distance from the skin. Journal of Applied Physiology. 2013;**114**(6): 778-785. DOI: 10.1152/japplphysiol.01271.2012

[13] Siddique ARM, Mahmud S, Heyst BV. A review of the state of the science on wearable thermoelectric power generators (TEGs) and their existing challenges. Renewable and Sustainable Energy Reviews. 2017;**73**: 730-744. DOI: 10.1016/j.rser.2017.01.177

[14] Lucia U. Carnot efficiency: Why? Physica A: Statistical Mechanics and its Applications. 2013;**392**(17):3513-3517. DOI: 10.1016/j.physa.2013.04.020

[15] Lay-Ekuakille A, Vendramin G, Trotta A, Mazzotta G. Thermoelectric

generator design based on power from body heat for biomedical autonomous devices. In: 2009 IEEE International Workshop on Medical Measurements and Applications. Cetraro, Italy: IEEE; 2009. pp. 1-4. DOI: 10.1109/MEMEA.2009.5167942

[16] Jeong MH, Kim KC, Kim JS, Choi KJ. Operation of wearable thermoelectric generators using dual sources of heat and light. Advancement of Science. 2022;**9**(12):1-9. DOI: 10.1002/advs.202104915

[17] Sattar M, Yeo WH. Recent advances in materials for wearable thermoelectric generators and biosensing devices. Materials. 2022;**15**(12):1-27. DOI: 10.3390/ma15124315

[18] Muchuweni E, Mombeshora ET. Enhanced thermoelectric performance by single-walled carbon nanotube composites for thermoelectric generators: A review. Applied Surface Science Advances. 2023;**13**. DOI: 10.1016/j.apsadv.2023.100379

[19] Shen D, Duley WW, Peng P, Xiao M, Feng J, Liu L, et al. Moisture-enabled electricity generation: From physics and materials to self-powered applications. Advanced Materials. 2020;**32**(52):1-31. DOI: 10.1002/adma.202003722

[20] Wei Q, Ge W, Yuan Z, Wang S, Lu C, Feng S, et al. Moisture electricity generation: Mechanisms, structures, and applications. Nano Research. 2023;**16**: 7496-7510. DOI: 10.1007/s12274-023-5465-9

[21] Enescu D. Models and indicators to assess thermal sensation under steady-state and transient conditions. Energies. 2019;**12**:841

[22] Hall JE. Energetics and metabolic rate. In: Hall JE, editor. Textbook of Medical Physiology. 13th ed. Philadelphia: Elsevier; 2016. pp. 903-907

[23] Francioso L, De Pascali C. Thermoelectric energy harvesting for powering wearable electronics. In: Dávila Pineda D, Rezania A, editors. Thermoelectric Energy Conversion: Basic Concepts and Device Applications. 1st ed. Weinheim: Wiley-VCH; 2017. pp. 205-231. DOI: 10.1002/9783527698110.ch10

[24] Houdas Y, Ring EFJ. Human Body Temperature: Its Measurement and Regulation. New York: Springer Science and Business Media; 2013. p. 238. DOI: 10.1007/978-1-4899-0345-7

[25] Çengel YA. Heat Transfer: A Practical Approach. 2nd ed. New York: McGraw-Hill; 2003. p. 978

[26] International Standardization Organization. Ergonomics of the Thermal Environment—Vocabulary and Symbols. 1st ed. Geneva: International Standardization Organization; 2001 Standard ISO 13731:2001

[27] Clark RP. Human skin temperature and convective heat loss. In: Cena K, Clark JA, editors. Bioengineering, Thermal Physiology, and Comfort. Amsterdam: Elsevier; 1981. p. 57. 288 p

[28] Dear RJ, Arens E, Hui Z, Oguro M. Convective and radiative heat transfer coefficients for individual human body segments. International Journal of Biometeorology. 1997;**40**:141-156. DOI: 10.1007/s004840050035

[29] Incropera FP, DeWitt DP, Bergman TL, Lavine AS. Fundamentals of Heat and Mass Transfer. 6th ed. Hoboken, NJ: Wiley; 2007. p. 1048

[30] Psikuta A, Kuklane K, Bogdan A, Havenith G, Annaheim S, Rossi RM.

Opportunities and constraints of presently used thermal manikins for thermo-physiological simulation of the human body. International Journal of Biometeorology. 2015;**60**:435-446. DOI: 10.1007/s00484-015-1032-6

[31] Havenith G, Holmér I, Parsons K. Personal factors in thermal comfort assessment: Clothing properties and metabolic heat production. Energy and Buildings. 2002;**34**(6):581-591. DOI: 10.1016/S0378-7788(02)00008-7

[32] Du Bois D, Du Bois EF. Clinical calorimetry: Tenth paper a formula to estimate the approximate surface area if height and weight be known. Archives of Internal Medicine. 1916;**17**:863-871. DOI: 10.1001/archinte.1916.00080130010002

[33] de Dear RJ. Global database of thermal comfort field experiments. In: Proceedings of the 1998 ASHRAE Winter Meeting; 18–21 January 1998; San Francisco, CA, USA. ASHRAE Transactions. Vol. 104, No. 1B. 1998. pp. 1141-1152

[34] Atmaca I, Kaynakli O, Yigit A. Effects of radiant temperature on thermal comfort. Building and Environment. 2007;**42**:3210-3220. DOI: 10.1016/j.buildenv.2006.08.009

[35] Enescu D. A review of thermal comfort models and indicators for indoor environments. Renewable and Sustainable Energy Reviews. 2017;**79**: 1353-1379. DOI: 10.1016/j.rser.2017.05.175

[36] Butera FM. Chapter 3—Principles of thermal comfort. In: Sayigh A, editor. Renewable and Sustainable Energy Reviews. Vol. 2, Issues 1–2. 1st ed. Amsterdam: Elsevier; 1998. pp. 39-66. DOI: 10.1016/S1364-0321(98)00011-2

[37] Choudhury AKR, Majumdar PK, Datta C. Chapter 1—Factors affecting comfort: Human physiology and the role of clothing. In: Guowen S, editor. Improving Comfort in Clothing. Woodhead Publishing Series in Textiles. Cambridge: Woodhead Publishing; 2011. pp. 3-60. DOI: 10.1533/9780857090645.1.3

[38] Zhang A, Li G, Wang B, Wang J. A theoretical model for wearable thermoelectric generators considering the effect of human skin. Journal of Electronic Materials. 2021;**50**:1514-1526. DOI: 10.1007/s11664-020-08695-6

[39] Wijethunge D, Kim D, Kim W. Simplified human thermoregulatory model for designing wearable thermoelectric devices. Journal of Physics D: Applied Physics. 2018;**51**: 055401. DOI: 10.1088/1361-6463/aaa17e

[40] Zhang A, Pang D, Lou J, Wang J, Huang WM. An analytical model for wearable thermoelectric generators harvesting body heat: An opportunistic approach. Applied Thermal Engineering. 2024;**236**:121658. DOI: 10.1016/j.applthermaleng.2023.121658

[41] Pang D, Zhang A, Guo Y, Wu J. Energy harvesting analysis of wearable thermoelectric generators integrated with human skin. Energy. 2023;**282**: 128850. DOI: 10.1016/j.energy.2023.128850

[42] Park H, Eom Y, Lee D, Kim J, Kim H, Park G, et al. High power output based on watch-strap-shaped body heat harvester using bulk thermoelectric materials. Energy. 2019;**187**:115935. DOI: 10.1016/j.energy.2019.115935

[43] Tabaie Z, Omidvar A. Human body heat-driven thermoelectric generators as a sustainable power supply for wearable

electronic devices: Recent advances, challenges, and future perspectives. Heliyon. 2023;**9**:14707. DOI: 10.1016/j.heliyon.2023.e14707

[44] Wang Y, Zhu P, Li W, Liu X, Li H, Deng Y, et al. High interfacial thermal stability of flexible flake-structured aluminum thin-film electrodes for Bi2Te3-based thermoelectric devices. ACS Applied Materials & Interfaces. 2022;**14**:12920-12926. DOI: 10.1021/acsami.2c00542

[45] Cao T, Shi X-L, Chen Z-G. Advances in the design and assembly of flexible thermoelectric devices. Progress in Materials Science. 2023;**131**:101003. DOI: 10.1016/j.pmatsci.2023.101003

[46] Jo SE, Kim MK, Kim MS, Kim YJ. Flexible thermoelectric generator for human body heat energy harvesting. Electronics Letters. 2012;**48**(16):1015-1017. DOI: 10.1049/el.2012.1566

[47] Sakamoto T, Iida T, Sekiguchi T, Taguchi Y, Hirayama N, Nishio K, et al. Selection and evaluation of thermal interface materials for reduction of the thermal contact resistance of thermoelectric generators. Journal of Electronic Materials. 2014;**43**:3792-3800. DOI: 10.1007/s11664-014-3165-7

[48] Wang Y, Shi Y, Mei D, Chen Z. Wearable thermoelectric generator for harvesting heat on the curved human wrist. Applied Energy. 2017;**205**:710-719. DOI: 10.1016/j.apenergy.2017.08.117

[49] Eom Y, Wijethunge D, Park H, Park SH, Kim W. Flexible thermoelectric power generation system based on rigid inorganic bulk materials. Applied Energy. 2017;**206**:649-656. DOI: 10.1016/j.apenergy.2017.08.076

[50] Kuang N, Niu A, Wang W, Zuo Z, Zhan T, Wang H. High-performance flexible thermoelectric generator using bulk legs and integrated electrodes for human energy harvesting. Energy Conversion and Management. 2022;**272**: 116337. DOI: 10.1016/j.enconman.2022.116337

[51] Khan F, Kim DH, Lee J. Functionalized materials and geometric designs of thermoelectric devices for smart wearable applications. Applied Energy. 2025;**379**:124940. DOI: 10.1016/j.apenergy.2024.124940

[52] Hu B, Shi X-L, Zou J, Chen Z-G. Thermoelectrics for medical applications: Progress, challenges, and perspectives. Chemical Engineering Journal. 2022;**437**:135268. DOI: 10.1016/j.cej.2022.135268

[53] Toshima N, Oshima K, Anno H, Nishinaka T, Ichikawa S, Iwata A, et al. Novel hybrid organic thermoelectric materials:Three-component hybrid films consisting of a nanoparticle polymer complex, carbon nanotubes, and vinyl polymer. Advanced Materials. 2015; **27**(13):2246-2251. DOI: 10.1002/adma.201405463

[54] Soleimani Z, Zoras S, Ceranic B, Shahzad S, Cui Y. A review on recent developments of thermoelectric materials for room-temperature applications. Sustainable Energy Technologies and Assessments. 2020;**37**: 100604. DOI: 10.1016/j.seta.2019.100604

[55] Shi H, Liu C, Jiang Q, Xu J. Effective approaches to improve the electrical conductivity of PEDOT:PSS: A review. Advanced Electronic Materials. 2015; **1**(4):1500017. DOI: 10.1002/aelm.201500017

[56] Jung KK, Jung Y, Choi CJ, Lee JM, Ko JS. Flexible thermoelectric generator with polydimethyl siloxane in

thermoelectric material and substrate. Current Applied Physics. 2016;**16**(10): 1442-1448. DOI: 10.1016/j.cap.2016.08.010

[57] Leonov V. Thermoelectric energy harvesters for powering wearable sensors. In: Sensors. Taipei, Taiwan: IEEE; 2012. pp. 1-4. DOI: 10.1109/ICSENS.2012.6411414

[58] Wu Q, Hu J. A novel design for a wearable thermoelectric generator based on 3D fabric structure. Smart Materials and Structures. 2017;**26**:045037. DOI: 10.1088/1361-665X/aa5694

[59] Chen J, He T, Du Z, Lee C. Review of textile-based wearable electronics: From the structure of the multi-level hierarchy textiles. Nano Energy. 2023;**117**:108898. DOI: 10.1016/j.nanoen.2023.108898

[60] Wu B, Li J, Tang J, Guo L, Lu Y, Wu Y, et al. Stretchable thermoelectric generators with enhanced output by infrared reflection for wearable application. Chemical Engineering Journal. 2023;**453**:139749. DOI: 10.1016/j.cej.2022.139749

Chapter 7

Controlling Heat Transfer through Walls and Roofs to Regulate Indoor Temperature: Analysis Using HT-Flux Software

Leonardus Murialdo Fransislus Purwanto, Hermawan and Stephanus Evert Indrawan

Abstract

Building design is currently recognized as a significant contributor to energy consumption, accounting for 24.79% of the total energy used by society. This consumption is projected to increase by 1% annually. Therefore, effective strategies for controlling heat transfer through walls and roofs must be considered and implemented in every building in humid tropical regions, such as Indonesia. This study consists of an introduction, research methods, literature review, object work analysis, and conclusions. This study aims to provide practical solutions by utilizing HT-Flux software to calculate heat transfer through walls and roofs in humid tropical areas, thereby minimizing its impact on indoor temperature increases. The research adopts a descriptive quantitative method, focusing on calculating heat transfer through walls and roofs using HT-Flux software based on finite element principles. The findings of this study serve as a guide for designing buildings in humid tropical regions that can reduce indoor air temperature increases, thereby lowering electricity consumption when air conditioning is used. By addressing this issue, the rise in electricity consumption can be effectively controlled and minimized. Key findings reveal that optimized wall and roof designs significantly reduce heat transfer, which, in turn, decreases indoor temperature fluctuations. These improvements result in a reduction in electricity consumption for air conditioning by up to 15–20%, depending on material selection and design modifications. The simulation also highlights the critical role of thermal insulation and reflective roofing materials in enhancing energy efficiency. This study provides practical guidance for designing energy-efficient buildings, ultimately contributing to sustainable energy management and mitigating the rise in electricity demand.

Keywords: HT-flux software, heat transfer, roof design, wall design, indoor air temperature

1. Introduction

Buildings play a vital role in human life, serving as spaces for living, working, and various public activities. However, buildings are also among the largest contributors to global energy consumption. A significant portion of this energy is used to maintain comfortable indoor temperatures, especially in regions with warm climates. One of the main factors influencing energy consumption in buildings is heat transfer through walls and roofs [1].

Uncontrolled heat transfer can lead to an increase in indoor temperature, forcing air conditioning systems to work harder to maintain comfort. This, in turn, requires higher electrical energy consumption. Besides increasing electricity usage, it also raises operational costs and exacerbates environmental impacts. Therefore, understanding and managing heat transfer through walls and roofs is essential for creating energy-efficient and environmentally friendly buildings [2].

Advancements in technology have made it easier to analyze heat transfer in buildings using software tools such as HT-Flux. This software enables accurate calculations based on finite element principles, providing detailed insights into how heat moves through wall and roof materials. Moreover, HT-Flux is simpler to use compared to other software, as research object designs can be imported directly from AutoCAD, and the calculation process is relatively faster despite using the same finite element model [3].

This study aims to analyze and propose solutions for controlling heat transfer through walls and roofs using HT-Flux software. The findings are expected to serve as a guide for building designers to create designs that minimize indoor temperature increases, reduce electricity consumption, and promote overall energy efficiency.

2. Research methodology

This study adopts a quantitative approach to analyze heat transfer through walls and roofs and to develop effective solutions for its regulation. The data utilized include the thermal properties of materials, such as thermal conductivity, specific heat capacity, and thickness, which were obtained from literature and technical material specifications. Environmental data, including external temperature, solar radiation, and humidity, were also incorporated as input parameters for the simulation.

The simulation process was conducted using HT-Flux software based on the finite element method. Wall and roof models were constructed, accounting for variations in materials, layers, and environmental conditions, to calculate temperature distribution and heat flux. The simulation results were quantitatively analyzed to evaluate the impact of heat transfer on indoor temperature and to assess the effectiveness of different designs and materials in reducing heat transfer.

Validation was carried out by comparing the simulation outcomes with empirical data or findings from previous studies to ensure reliability. Based on the analysis, recommendations for thermally efficient wall and roof designs were formulated, emphasizing the reduction of indoor temperature fluctuations and energy consumption for cooling purposes. This approach aims to provide practical solutions for sustainable building design.

3. Literature review

The control of heat transfer through walls and roofs is crucial in designing energy-efficient buildings. Uncontrolled heat transfer can significantly increase the demand for cooling or heating, thus escalating energy consumption. Simulation tools such as HT-Flux provide an in-depth analysis of these processes, assisting in the creation of energy-efficient building designs [4]. Heat transfer occurs through three primary mechanisms: conduction, convection, and radiation. Conduction refers to the flow of heat through solid materials, convection pertains to heat transfer *via* fluids such as air, and radiation involves the transfer of energy through electromagnetic waves. Materials with low thermal conductivity, such as polyurethane-based or mineral fiber insulation, effectively minimize heat transfer through walls and roofs [5]. Reflective coatings applied to roofs are also instrumental in reducing thermal radiation [6]. Architectural design significantly influences heat transfer control. Elements such as window placement, building orientation, and the incorporation of green roofs are pivotal strategies. Green roofs, for example, have been shown to reduce surface temperatures substantially, thereby reducing the cooling load in buildings as low as 2.7°C [7, 8]. Advances in building materials, such as the utilization of Phase Change Materials (PCM), which absorb and release thermal energy at specific temperatures, also contribute to maintaining stable indoor temperatures [9]. Furthermore, nano-technology-derived materials like aerogel offer exceptionally low thermal conductivity with minimal material thickness, making them highly efficient thermal insulators [10]. HT-Flux is a software tool designed to simulate heat transfer in building structures. Utilizing the finite element method, HT-Flux provides a detailed analysis of the temperature distribution and heat flux in building components, especially in the depiction of heat distribution that cannot be done with manual calculations. Studies confirm that HT-Flux delivers accurate simulations when compared to field measurements, with an average deviation of less than 5% [11]. In building design, HT-Flux is employed to evaluate the thermal performance of new materials, assess the effectiveness of insulation and ventilation strategies, and identify hotspots within buildings for further optimization [12]. Several case studies highlight the effectiveness of HT-Flux in achieving energy-efficient building designs. For instance, simulations using HT-Flux have shown that PCM-based wall insulation significantly reduces heat flux, lowering cooling energy requirements [13]. Similarly, reflective roof coatings have been proven to decrease indoor temperatures during summer months. However, challenges remain, such as the high costs of advanced thermal insulation materials and emerging technologies like PCM and aerogel. Moreover, the durability of these materials under extreme environmental conditions continues to be an active area of research. Future prospects include the integration of thermal simulations with artificial intelligence to optimize building designs automatically. Additionally, leveraging real-time data from thermal sensors can enhance simulation accuracy and allow for dynamic adaptation to environmental changes. In conclusion, the control of heat transfer through walls and roofs is essential for improving building energy efficiency. Software tools like HT-Flux provide a robust platform for analyzing and optimizing heat transfer, supporting the development of sustainable and energy-efficient building designs.

4. Measurement of research objects

The buildings observed in this study were selected based on the following structural variations:

1. Walls constructed with half-brick thickness (15 cm).

2. Walls constructed with full-brick thickness (30 cm).

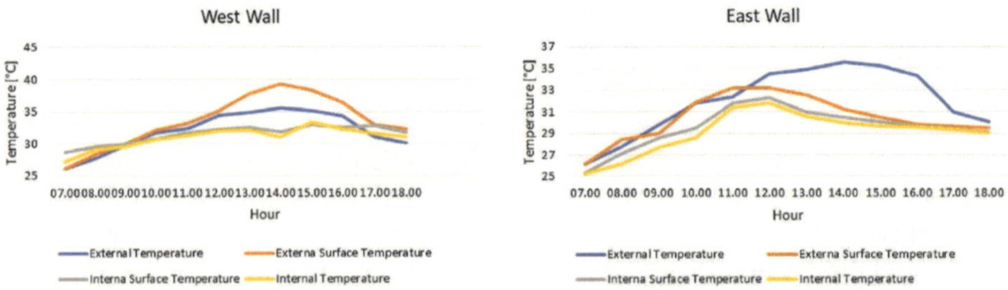

Figure 1.
Outdoor air temperature, exterior wall surface temperature, interior wall surface temperature, and indoor air temperature for the east and west walls.

3. Ground-floor walls.
4. First-floor walls.
5. Walls with glass windows.
6. Walls with wooden windows.
7. Unplasticized polyvinyl chloride (UPVC) roofs without ceilings.
8. UPVC roofs with ceilings.
9. Clay tile roofs without ceilings.
10. Clay tile roofs with ceilings.

The dimensions and types of materials are those that are often used in Indonesia, which has a humid tropical climate and has strong sun exposure.

For the measurement process, the following instrument was used:

1. A digital thermometer/hygrometer (Lutron HT3004) to measure air temperature. For humidity measurement range with a range of 10–95% and a temperature measurement range of 0–60°V. This tool has been calibrated by the Lutron vendor
2. Digital Doppelthermometer Voltcraft 302 K/J for wall surface temperature measurement. The measurement of the wall surface temperature is around 0 to 250°C. This tool has been calibrated by the Voltcraft vendor

The measurement results are presented in the following diagram (see **Figure 1**).

5. Starting the simulation with HT-flux software

The HT-Flux software, developed by HT-Flux Engineering GmbH in Austria, is capable of calculating heat flow, U-value, Ψ value, fRsi value, and extreme temperature value, all of which are used in heat transfer calculations. These calculations are essential for analyzing heat transfer through walls and roofs, which significantly

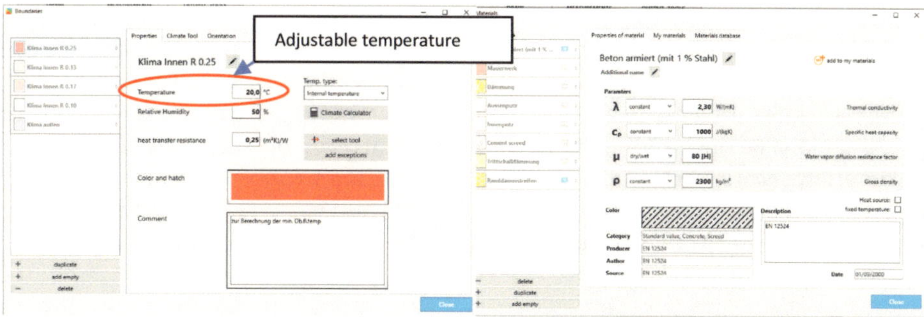

Figure 2.
Data input in HT-Flux that can be set up with local climate data adjustments.

impact thermal comfort within buildings and efforts to optimize energy usage in tropical regions for artificial conditioning. Heat leakage into the building increases indoor temperatures, leading to higher energy consumption to reduce these temperatures.

This program provides features for adding materials with data inputs for λ (thermal conductivity), μ (water vapor diffusion resistance factor), ρ (density), and Cp (specific heat capacity). Additionally, external and internal temperature data can be modified, including inputs for temperature, humidity, heat transfer resistance, and the orientation of the measured object. This change in temperature affects the heat transfer output because the temperature is adjusted to the local climate of each place, for example, in humid tropical areas that are different from those in subtropical climates. This input data can be arranged as in **Figure 2**.

6. Calculations on various wall shapes with HT-flux analysis

In heat transfer calculations, strict attention must be given to the U-value. The lower the U-value in a building's construction, the greater the influence of external heat entering the building. A U-value calculation involves determining the overall thermal transmittance of a wall, accounting for the thermal resistances of its layers. For instance, consider a wall with three layers: an exterior brick layer (thermal resistance 0.25 m^2K/W), insulation (1.50 m^2K/W), and interior plaster (0.10 m^2K/W). The total thermal resistance (R_{total}) is the sum of individual resistances:

$R_{total} = 0.25 + 1.50 + 0.10 = 1.85\ m^2K/W$. The U-value is the reciprocal of the total resistance: U = 1/R_total = 1/1.85 ≈ 0.54 W/m^2K. This value indicates the rate of heat transfer through the wall, emphasizing its insulation performance. However, calculations related to thermal bridges must also be considered, as they are often neglected in tropical regions. Thermal bridges signify significant heat leakage from the exterior to the interior of a building, which substantially affects indoor comfort. Thermal bridges are indeed critical points of heat transfer that can impact indoor comfort and energy efficiency. In our study, we explored specific strategies to mitigate thermal bridging, including (1) incorporating continuous insulation to minimize gaps in the building envelope, (2) using thermally broken materials at junctions and connections, such as windows and door frames, (3) designing with optimized structural details to reduce conductive pathways, and (4) applying advanced insulating materials, such as aerogels, in challenging areas. These strategies aim to enhance thermal performance and ensure a more consistent indoor environment.

Thermal comfort calculations in tropical regions are frequently overlooked because the issue of heat often only results in unmet thermal comfort, and the pragmatic solution tends to involve installing air conditioning systems without considering the associated increase in energy consumption. This is particularly concerning in tropical regions where buildings are exposed to high levels of solar radiation.

This study tests several wall and roof configurations to observe heat transfer and its effects on indoor spaces. Calculations are conducted using HT-Flux software to analyze temperature gradients and thermal bridges occurring in various building constructions, as follows:

6.1 Half-brick wall

A half-brick wall with a thickness of 15 cm, inner and outer plaster layers of 1.5 cm each, and a 10 cm brick core were evaluated by inputting external and internal air temperatures. The temperature distribution and conductive heat flux in this wall type show that it has a limited ability to hinder heat transfer into the building.

The wall under analysis consists of a 15 cm thick half-brick layer as the main material, complemented by inner and outer plaster layers, each 1.5 cm thick, and an additional 10 cm brick layer. This wall structure is designed to reduce heat transfer between the external environment and the indoor space.

Heat transfer through the wall occurs due to the temperature difference between the outside air and the indoor environment. Materials like bricks have moderate thermal conductivity, approximately 0.72 W/mK, which allows heat to pass through at a controlled rate. The plaster layers, although thinner, have a thermal conductivity of about 0.5 W/mK [14], further contributing to the thermal resistance. The total wall thickness, approximately 28 cm, provides sufficient thermal resistance to maintain indoor temperature stability. The plaster layers, while thinner compared to the brick layer, contribute to the overall thermal resistance of the wall by providing an additional barrier to heat transfer. With a thermal conductivity of approximately 0.5 W/mK and a typical thickness of around 10–20 mm, these layers have a relatively small individual resistance. However, when combined with other materials in the wall assembly, they play a cumulative role in reducing heat flow. By slowing down the rate of heat transfer, the plaster layers enhance the overall thermal performance of the wall, complementing the insulating properties of the thicker brick layer.

As shown in the simulation (as seen in **Figure 3**), the temperature distribution across the wall forms a gradient, with higher temperatures on the outside gradually decreasing toward the inside. This indicates that the wall effectively reduces heat flow from the external environment into the indoor space. With this structure, indoor temperatures can remain stable even when outdoor temperatures fluctuate.

The thermal efficiency of this wall is particularly beneficial for enhancing indoor thermal comfort and reducing energy consumption for heating or cooling [15]. To further improve efficiency, additional insulation layers could be considered, especially in environments with significant differences between outdoor and indoor temperatures [16].

6.2 Full-Brick Wall

The total thickness of the full-brick wall is 30 cm, comprising 1.5 cm outer and inner plaster layers and a 27 cm brick core. This wall construction demonstrates increased thermal resistance and effectively reduces heat transfer through the wall.

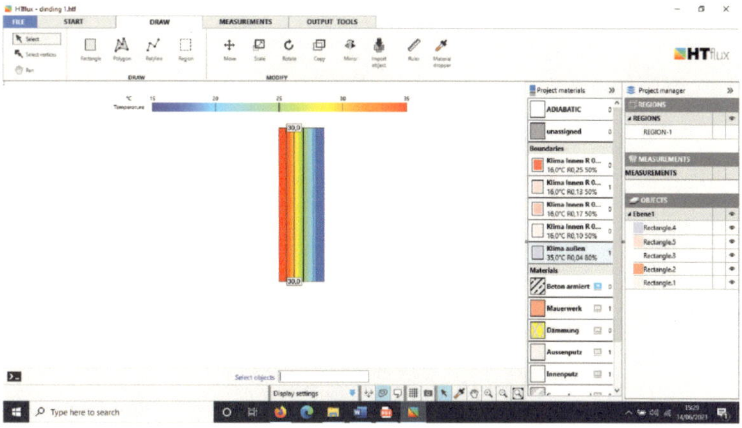

Figure 3.
Heat flux distribution on a half-brick wall.

The wall structure in question consists of a total thickness of 30 cm, which includes an outer plaster layer (1.5 cm), an inner plaster layer (1.5 cm), and a brick layer (27 cm). This configuration represents a single-brick wall, commonly used in construction for thermal and structural efficiency.

Heat transfer occurs due to the temperature difference between the external environment and the indoor space. The materials used in this wall have distinct thermal properties:

- Brick is the primary material, with a thermal conductivity of approximately 0.72 W/mK, known for its moderate ability to conduct heat [14].
- Plaster, though thinner, has a thermal conductivity of around 0.5 W/mK, contributing additional resistance to heat flow [15].

The simulation, as illustrated in the accompanying image, shows the temperature gradient across the wall, as seen in **Figure 4**. Higher temperatures on the outer surface gradually decrease toward the inner surface, indicating that the wall effectively reduces heat transfer from the external to the internal environment. This reduction in heat transfer helps maintain a more stable indoor temperature.

The thermal resistance of the wall is enhanced by the combination of materials and their respective thicknesses. While the brick layer provides the majority of the wall's structural and thermal performance, the plaster layers play a significant role in minimizing thermal bridging and enhancing the overall resistance to heat flow.

This wall structure is efficient for moderate climate conditions, as it offers a balance between structural strength and thermal performance. However, in extreme climates, additional insulation layers could further improve its energy efficiency by reducing heating or cooling loads [18]. Such enhancements would align with modern energy-saving practices and provide long-term cost benefits.

6.3 Half-brick wall with wooden frames and glass windows

Heat transfer through glass significantly contributes to heat penetration from outside to inside the building. Proper sealing at the junctions between wooden frames and

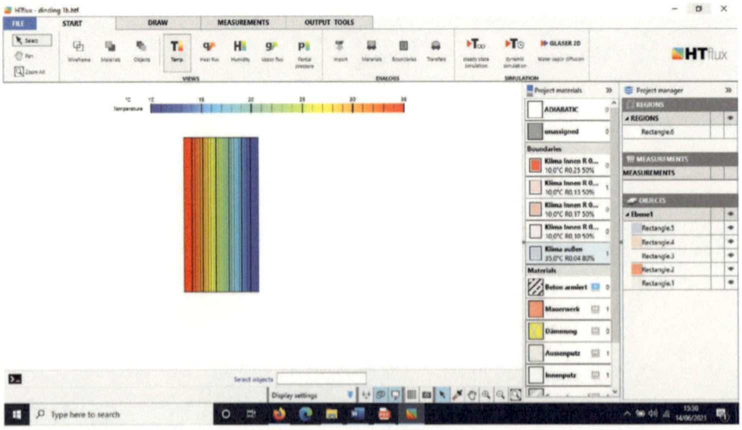

Figure 4.
Heat flux distribution on a full-brick wall.

brick walls can prevent heat leakage. However, to minimize heat transfer through the connection between glass and frames, sealants must be applied to prevent heat ingress.

This study investigates the heat transfer characteristics of a wall system comprising a half-brick wall with integrated wooden frames and glass windows, as illustrated in the provided simulation image as in **Figure 5**. The analysis evaluates how different materials in the wall system contribute to heat transfer, focusing on their thermal properties and their impact on overall energy efficiency.

The primary component of the wall is a half-brick layer, approximately 12.5 cm thick. Bricks are known for their moderate thermal conductivity, typically ranging from 0.60 to 0.75 W/mK, depending on their density and composition [16]. In this configuration, the brick wall acts as a thermal barrier, slowing the heat transfer from the external environment to the indoor space. However, its performance can be affected by the presence of thermal bridges, such as wooden frames and glass windows.

Wooden frames, which are commonly used for structural and esthetic purposes, exhibit lower thermal conductivity compared to bricks, with values typically around 0.15 W/mK [17]. While wood provides some resistance to heat transfer, it can create

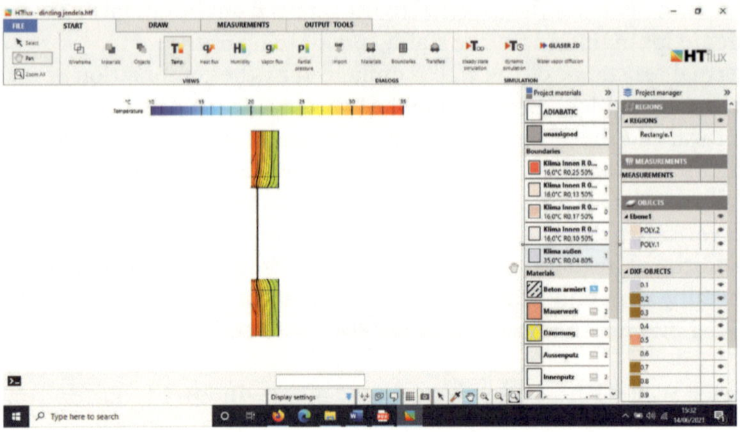

Figure 5.
Heat flux distribution on a wall with wooden frames and glass windows.

thermal bridges due to its direct contact with both the external and internal surfaces. These thermal bridges may allow localized heat transfer, slightly reducing the thermal performance of the overall wall system.

The glass windows in this configuration are key elements influencing heat transfer. Glass typically has a higher thermal conductivity than both bricks and wood, approximately 0.8–1.0 W/mK [18]. This makes windows a potential weak point in the wall system for resisting heat flow. However, the inclusion of modern glazing technologies, such as double-glazed or low-emissivity (low-E) glass, can significantly improve thermal insulation by trapping air or gas layers between glass panes, thereby reducing heat transfer. The effectiveness of these technologies depends on the specific type of glass and coating used.

The temperature distribution, as shown in the HT-Flux simulation, reveals the interaction between these materials. Higher temperatures on the external surface gradually transition to lower temperatures inside the building. The presence of wooden frames and glass windows introduces localized variations in the temperature gradient, highlighting the need for careful material selection and design to optimize thermal performance.

This wall system's thermal performance is suitable for moderate climates but could be enhanced for extreme conditions. Adding insulating materials around wooden frames and utilizing advanced glazing solutions for windows are effective measures to reduce heat loss or gain. Such improvements can significantly reduce energy consumption for heating or cooling, aligning with sustainability goals in modern construction [19].

To further optimize the thermal performance of similar wall systems, the following strategies are recommended:

- Insulation: Integrating insulation materials, such as polyurethane or mineral wool, within the brick wall or around wooden frames.

- Advanced Glazing: Adopting triple-glazed or low-E glass to minimize heat transfer through windows.

- Sealing: Ensuring airtight sealing between the wooden frames and brick wall to prevent air leakage, which can compromise energy efficiency.

6.4 Second-floor brick wall meeting a concrete floor slab

At the junction between a second-floor brick wall and a concrete floor slab, proper adhesion must be ensured to prevent heat transfer into the building. **Figure 6** illustrates the deflection of isotherm lines at the wall-slab interface due to differences in thermal resistance and highlights the critical areas at the junction that require attention to prevent heat leakage.

This analysis examines the heat transfer characteristics in a two-story brick wall structure with a concrete slab intersection, as depicted in the provided simulation. The study focuses on the thermal interaction between brick walls and concrete, evaluating their roles in heat transfer and the implications for energy efficiency.

The brick wall, with a typical thermal conductivity of 0.60–0.75 W/mK, serves as the primary thermal barrier for the building envelope [20]. Its moderate conductivity allows it to resist heat flow effectively, maintaining a stable indoor temperature. The concrete slab, on the other hand, has a higher thermal conductivity, ranging from 1.2 to

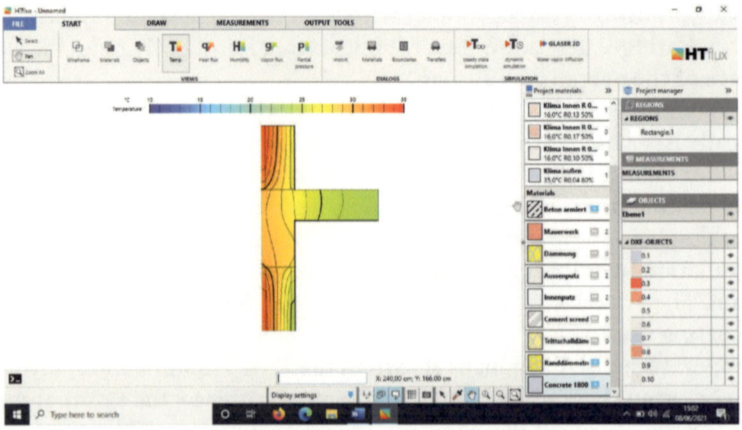

Figure 6.
Heat flux distribution at the second-floor wall and concrete floor slab interface.

1.8 W/mK, depending on its density and aggregate composition [21]. This higher conductivity makes the slab a potential thermal bridge at the junction with the brick wall.

The intersection between the brick wall and the concrete slab is a critical point for heat transfer. Concrete's higher thermal conductivity facilitates heat flow, potentially compromising the overall thermal resistance of the wall system. This effect is exacerbated if the intersection is not properly insulated, leading to localized heat loss or gain. The HT-Flux simulation shows temperature gradients indicating heat flow through the wall-slab interface (in **Figure 6**). These gradients highlight the importance of thermal breaks or insulation at the junction to mitigate heat transfer. Materials like extruded polystyrene or aerogels can be used to create a thermal barrier at this point [22].

Addressing the thermal bridging at the wall-slab interface is essential for energy efficiency. Proper insulation at the junction can significantly reduce heat loss, improving indoor thermal comfort and reducing energy consumption for heating or cooling. For optimal results, the design should also ensure airtight construction to minimize air leakage.

6.5 Ground-floor brick wall

Similar to the second-floor wall junction, attention should be given to the connection between the wall and the floor on the ground floor. The deflection of isotherm lines at the junction marks a weak point. Proper adhesion systems and mortar must be employed to prevent heat leakage from outside into the building. The deflection of isotherm lines at a junction indicates areas of uneven heat flow, which can signify thermal bridges. These weak points disrupt the uniform distribution of temperature across the building envelope, leading to increased heat loss in colder conditions or unwanted heat gain in warmer conditions. As a result, the building's thermal efficiency is compromised, requiring more energy to maintain indoor comfort. Additionally, such deflections can cause localized temperature drops, potentially leading to condensation and moisture-related issues. Addressing these weak points is essential to improving thermal performance and reducing energy consumption.

The thermal performance of a ground-floor brick wall is a critical factor in regulating indoor temperatures and ensuring energy efficiency. The wall in this

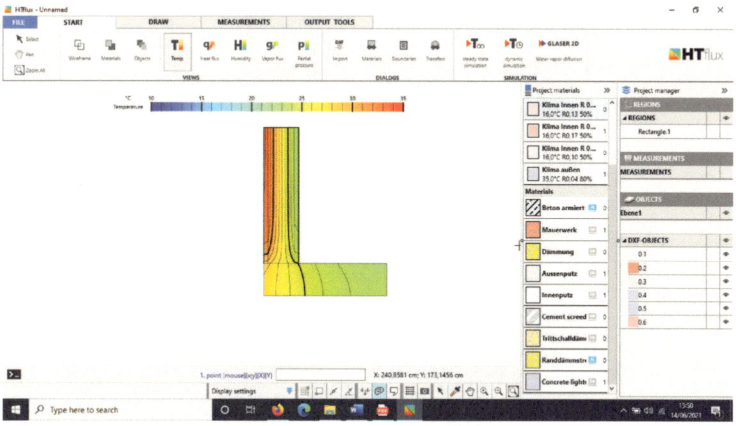

Figure 7.
Heat flux distribution at the ground-floor wall.

analysis consists of standard bricks with an approximate thermal conductivity of 0.60–0.75 W/mK [23]. This material offers moderate resistance to heat transfer, providing a balance between structural integrity and thermal insulation.

Ground-floor walls are subject to unique thermal dynamics due to their interaction with both external air and the foundation. Heat transfer in such walls occurs primarily *via* conduction through the brick material and convection at the surfaces. The interface between the wall and the ground presents a potential thermal bridge, as the soil or foundation materials often have varying thermal properties [24]. To minimize thermal bridges at the wall-ground interface, it is recommended to use foundation materials with low thermal conductivity, such as extruded polystyrene (XPS) or rigid foam insulation. These materials provide effective insulation, reducing heat transfer and improving overall thermal performance at the junction. The HT-Flux simulation illustrates the temperature gradients across the wall, showing how heat is conducted from the exterior to the interior (as in **Figure 7**). The analysis highlights the importance of minimizing heat transfer at the wall's base, where ground contact can lead to thermal losses. Incorporating a layer of insulation, such as rigid foam boards or polyurethane, at the base can significantly enhance thermal resistance and reduce energy consumption.

Optimizing ground-floor brick walls involves selecting appropriate materials and addressing thermal bridging at the foundation. These measures improve indoor comfort and contribute to energy efficiency in buildings.

6.6 Clay tile roof and wall construction

In the construction of clay tile roofs and walls, a uniform heat distribution is observed across the roof, with the upper side of the roof showing higher temperatures compared to the underside. The heat flux diagram at the roof-wall junction highlights areas of significant heat distribution and heat ingress through these joints. The heat effect on the walls influences heat propagation through the wooden structures at the edges of the roof construction.

In this construction, as shown in **Figure 8**, all areas exhibit elevated temperatures. The heat influence from the roof is particularly dominant, affecting the entire structure due to high solar radiation exposure in tropical regions.

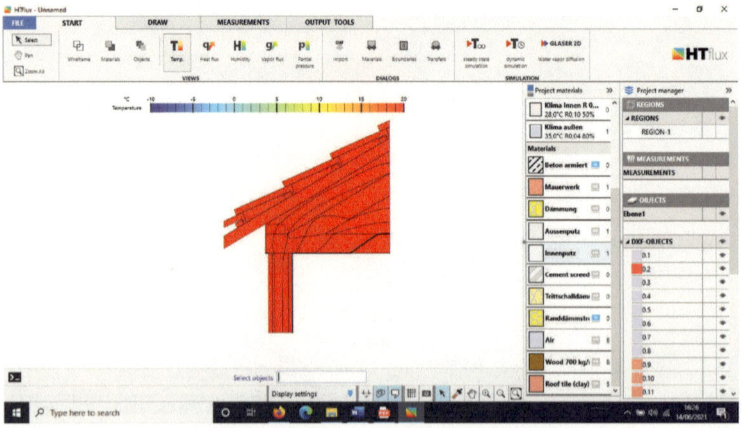

Figure 8.
Heat flux distribution in clay tile roof and wall construction.

6.7 UPVC roof and wall construction

UPVC roofing typically has a lower slope angle compared to clay tile roof constructions. This results in a more uniform distribution of heat across the roof, with significantly high temperatures throughout the structure.

Controlling heat transfer through walls and roofs is a crucial aspect of designing energy-efficient buildings. The use of materials such as UPVC (Unplasticized Polyvinyl Chloride) for walls and roofs offers advantages in thermal insulation. UPVC's low thermal conductivity helps reduce heat transfer between the external and internal environments of a building. Thermal analysis using software like HT-Flux, as depicted in the simulation image, provides a deeper understanding of temperature distribution in building elements as shown in **Figure 9**. The simulation results demonstrate that utilizing UPVC walls and roofs significantly impacts indoor temperatures, especially in tropical environments with high solar radiation. HT-Flux facilitates the visualization of how these materials act as effective thermal barriers, maintaining indoor temperature stability [25]. The application of UPVC also supports energy

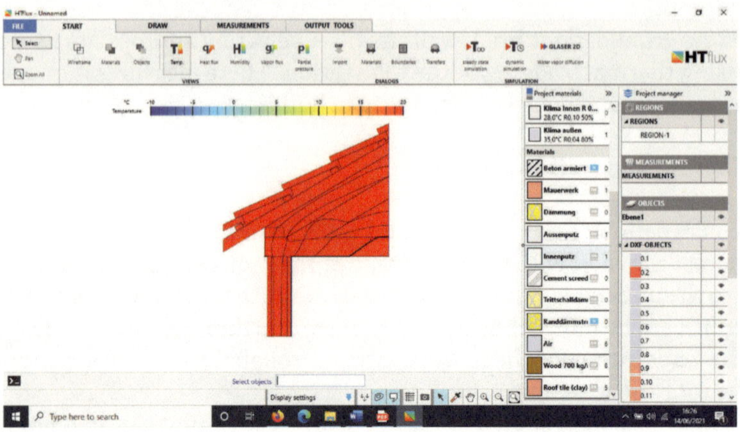

Figure 9.
Heat flux distribution in UPVC roof and wall construction.

efficiency by reducing the load on air conditioning systems. This study highlights essential insights for architects and engineers in designing sustainable and energy-efficient buildings. A detailed understanding of heat distribution through simulation enables the optimization of designs tailored to specific environmental conditions.

6.8 Clay tile roof and wall construction with gypsum ceiling

The addition of a gypsum ceiling beneath the roof construction significantly reduces indoor heat. This ceiling acts as an insulator, slightly lowering temperatures beneath the roof and creating a differentiated heat distribution within the roof structure. However, the roof-wall junction remains a critical area requiring thermal insulation to prevent heat ingress into the building.

The selection of materials for walls, roofs, and ceilings plays a critical role in controlling heat transfer and maintaining indoor thermal comfort. Clay roof tiles, often used in traditional and modern construction, provide excellent thermal insulation due to their natural composition and high thermal mass. When combined with a gypsum ceiling, which is lightweight and has low thermal conductivity, the system effectively reduces heat transfer into the interior space, as shown in the HT-Flux simulation [26].

HT-Flux, a specialized software for thermal analysis, enables the visualization of temperature distribution across building elements (as in **Figure 10**). The simulation results demonstrate the effectiveness of clay roof tiles in dissipating heat and the role of the gypsum ceiling in further insulating the indoor environment. This combination is particularly beneficial in regions with high solar radiation, ensuring indoor thermal comfort and reducing reliance on air conditioning systems.

The integration of clay tiles and gypsum ceilings not only supports energy efficiency but also enhances the durability and sustainability of the building. This study highlights the importance of selecting materials with complementary thermal properties to optimize building performance, particularly in tropical climates.

6.9 UPVC roof and wall construction with gypsum ceiling

After installing a gypsum ceiling, heat distribution becomes more uniform, and a reduction in roof temperature is observed. However, the lower slope angle of the

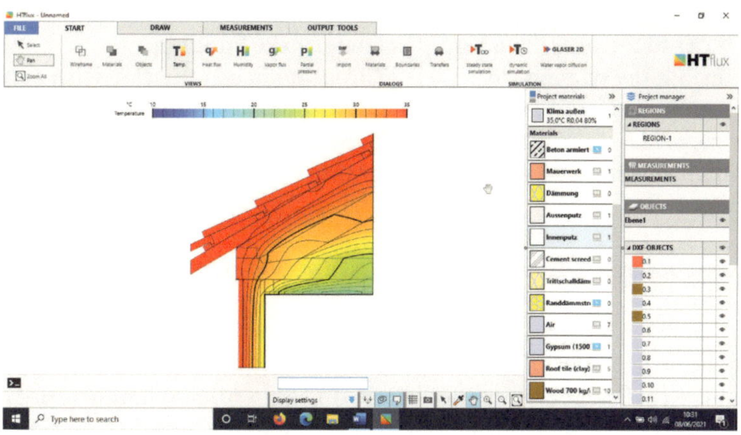

Figure 10.
Heat flux distribution in clay tile roof and wall construction with gypsum ceiling.

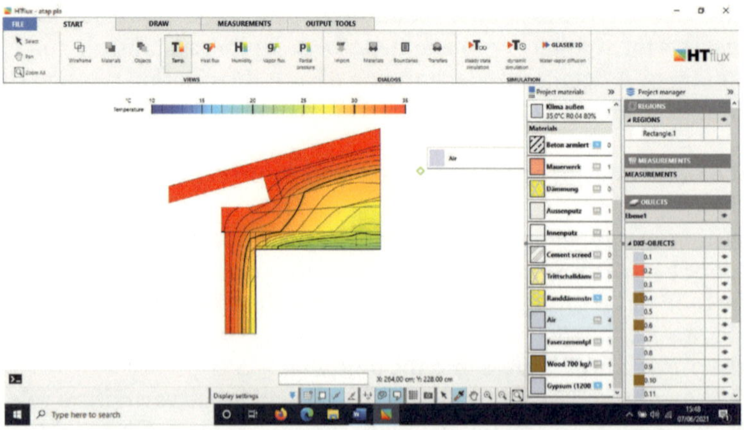

Figure 11.
Heat flux distribution in UPVC roof and wall construction with gypsum ceiling.

UPVC roof limits significant heat reduction, as the smaller air cavity beneath the roof provides less insulation and is unable to effectively resist heat transfer.

The combination of UPVC (unplasticized polyvinyl chloride) roofing and gypsum ceilings offers significant advantages in thermal insulation for buildings. UPVC roofing sheets exhibit low thermal conductivity, effectively reducing heat transfer and maintaining cooler indoor temperatures during hot weather conditions. This characteristic minimizes the reliance on air conditioning systems, leading to energy savings [27].

Gypsum ceilings further enhance thermal performance due to their inherent heat-resistant properties. The non-combustible core of gypsum boards contains chemically combined water, which, when exposed to high heat, is slowly released as steam, effectively retarding heat transfer. This mechanism contributes to maintaining indoor thermal comfort by acting as a barrier against external heat [28].

Utilizing HT-Flux software for thermal analysis allows for precise simulation of temperature distribution across building elements, as shown in **Figure 11**. The integration of UPVC roofing and gypsum ceilings, as depicted in the simulation, demonstrates a synergistic effect in reducing heat ingress, particularly beneficial in regions with high solar radiation. This combination not only enhances occupant comfort but also contributes to the overall energy efficiency of the building.

In conclusion, the strategic use of UPVC roofing in conjunction with gypsum ceilings presents an effective solution for controlling heat transfer in buildings. Employing tools like HT-Flux for detailed thermal analysis aids architects and engineers in optimizing building designs for improved energy performance and sustainability.

6.10 Clay tile roof and wall construction with gypsum ceiling and roof thermal insulation

Adding thermal insulation materials, such as Styrofoam and glass wool, within the air cavity of the roof structure has not resulted in a significant temperature reduction in either clay tile or asbestos roofing. This is evident in the heat distribution depicted in **Figure 12**. The roof-wall junction remains the area with the highest heat transfer, emphasizing the need for enhanced insulation in these critical junctions.

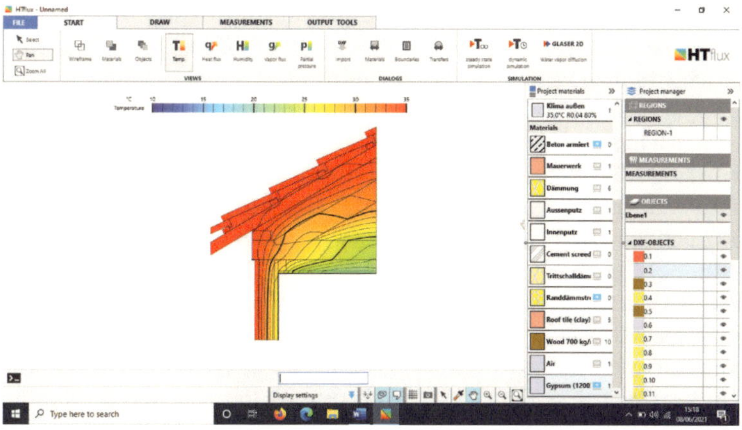

Figure 12.
Heat flux distribution in clay tile roof and wall construction with gypsum ceiling and roof insulation.

The combination of clay roof tiles, gypsum ceilings, and thermal insulation in roofs presents an effective solution for controlling heat transfer in buildings. Clay roof tiles are known for their high thermal mass, which helps dissipate heat from solar radiation and reduces its transmission to the interior. Adding a gypsum ceiling, which is lightweight and has excellent thermal resistance, further enhances the insulation. To improve this system, the inclusion of thermal insulation beneath the roof tiles acts as an additional barrier, effectively reducing heat transfer and ensuring indoor comfort [29].

HT-Flux software simulations, as illustrated, highlight the benefits of this material configuration. The clay tiles dissipate heat efficiently, while the thermal insulation and gypsum ceiling prevent heat from penetrating the living space. This setup is particularly beneficial in tropical and arid regions, where managing solar heat gain is critical. By integrating roof insulation, cooling loads can be significantly reduced, contributing to lower energy consumption and improved thermal comfort [30].

This study emphasizes the importance of combining materials with complementary thermal properties. By leveraging advanced thermal simulation tools like HT-Flux, architects, and engineers can optimize building designs for energy efficiency and sustainability, addressing the growing demand for climate-responsive construction solutions.

7. Conclusion

This research has demonstrated the critical role of walls and roofs in controlling heat transfer and regulating indoor temperatures, utilizing HT-Flux software for analysis. The strength of HT-Flux lies in its rapid calculations, ease of integrating temperature and material data inputs, and the clarity of its output, which makes the results more comprehensible. The findings emphasize the importance of material selection, design configurations, and thermal properties in ensuring effective thermal performance of building envelopes. By analyzing various material layers, thicknesses, and thermal gradients, this study provides insights into optimizing heat flow control to achieve energy efficiency and indoor comfort.

A key insight from this analysis is the significance of the thermal resistance of materials. Walls and roofs with higher thermal resistance slow down the rate of heat transfer, maintaining stable indoor temperatures despite external fluctuations. For example, multi-layered walls composed of plaster and bricks, as well as insulated roofs, effectively reduce heat flow, thereby minimizing energy demands for cooling during the dry season and heating during the wet season in humid tropical regions.

The study also highlights the role of thermal gradients in evaluating heat transfer. Simulations using HT-Flux software enable the visualization of temperature changes across various wall and roof layers. These gradients provide valuable information for designing construction elements that resist external heat flow while maintaining the desired indoor thermal conditions. Proper material combinations, such as dense bricks paired with insulation, help create a barrier that reduces heat transmission and optimizes energy efficiency.

Moreover, the findings underline the importance of layer thickness in thermal performance. Increasing the thickness of materials with low thermal conductivity, such as insulation or plaster, significantly enhances the ability of walls or roofs to resist heat transfer by up to 7% compared to unprotected counterparts. Conversely, thinner layers or the absence of insulation compromise the thermal efficiency of the system, leading to a 15–17% increase in energy consumption and discomfort in maintaining a stable indoor environment.

Roof design strategies also emerge as a critical aspect of heat transfer control. Roofs, being directly exposed to solar radiation, require special attention to their materials and structural design. The application of reflective coatings, insulation, or green roof systems can drastically reduce heat absorption, minimizing heat transfer into the building. These solutions are particularly crucial in tropical and arid climates, where roofs account for a significant portion of heat gain.

Additionally, the study emphasizes the need for dynamic thermal performance evaluation. The use of HT-Flux software provides a clear understanding of how various construction scenarios respond to changing external conditions. This approach allows designers and engineers to simulate real-world conditions, identify weaknesses in thermal envelopes, and make data-driven decisions to enhance performance. Visualizing heat flow and thermal gradients adds another layer of precision in evaluating and improving building designs.

The practical applications of these findings are extensive. Developers and architects can leverage insights from HT-Flux simulations to design buildings with optimized energy consumption and thermal comfort. Homeowners and facility managers can implement retrofitting measures, such as adding insulation or reflective coatings, to improve the performance of existing structures. Policymakers can use this data to establish building codes and standards that promote energy-efficient construction practices, ultimately contributing to sustainability goals.

Despite its clear benefits, this study acknowledges limitations in its scope. The simulations were conducted under controlled conditions, which may not fully account for variables such as moisture ingress, air leakage, or material degradation over time. Future research could expand on these aspects to provide a more comprehensive understanding of real-world performance. Additionally, integrating other software tools or experimental validation could strengthen the conclusions and broaden the applicability of the findings.

In conclusion, controlling heat transfer through walls and roofs is fundamental to regulating indoor temperatures and achieving energy efficiency. The analysis using HT-Flux software has demonstrated how material selection, layer configuration, and design strategies can synergistically enhance the thermal performance of building envelopes. By adopting these principles, stakeholders in the construction and design industry can create sustainable, energy-efficient buildings that meet modern comfort standards while minimizing environmental impact. As the demand for energy-efficient solutions grows, the insights from this study serve as a valuable resource for advancing sustainable construction practices and improving the overall quality of the built environment.

Author details

Leonardus Murialdo Fransislus Purwanto*, Hermawan and Stephanus Evert Indrawan
Soegijapranata Catholic University, Semarang, Indonesia

*Address all correspondence to: lmf_purwanto@unika.ac.id

References

[1] Soto FJZ et al. Energy consumption in buildings: A compilation of current studies. Energy Reports. 2025;**13**:1293-1307. DOI: 10.1016/j.egyr.2024.12.069

[2] Jaworski M. Thermal performance of building element containing phase change material (PCM) integrated with ventilation system – An experimental study. Applied Thermal Engineering. 2014;**70**(1):665-674. DOI: 10.1016/j.applthermaleng.2014.05.093

[3] Aria ZADP, Purwanto LMF. Comparison of heat transfer of wall materials with thermal simulation. Jurnal Arsitektur ARCADE. 2021;**5**(1):77-81. Available from: https://e-journal.ukri.ac.id/index.php/arcade/article/view/654

[4] Purwanto LMF. Heat transfer on residential roofs with the HT-Flux software. Journal ARTEKS. 2023;**8**(3):495-502. DOI: 10.30822/arteks.v8i3.2775

[5] HTflux – The Smart Building Physics Software. HTflux. [Online]. Available from: https://www.htflux.com/en

[6] Paulus DR, Purwanto LMF. Comparison of heat transfer in software Psi-Therm, HT-Flux and Ansys. Journal of Digital Architecture. 2022;**1**(2):98-109. DOI: 10.24167/joda.v1i2.4306

[7] Heat-Flux Measurement Tool. HTflux. [Online]. Available from: https://www.htflux.com/en/documentation/measuring-tools/heat-flux-measurement

[8] Pandora N. The Influence of Green Roof and Façade Design against Thermal Comfort at the Tanatap Coffee Meruya Building Jakarta, Research Report Book. Bandung: Parahyangan Catholic University

[9] Purwanto LMF. Simulation of heat transfer on walls with Therm 7.7 software. Journal ARTEKS. 2019;**4**(1):111-115. DOI: 10.30822/arteks.v4i1.215

[10] Installing HTflux. HTflux. [Online]. Available from: https://www.htflux.com/en/documentation/installing-htflux

[11] Effendi AC, Purwanto LMF. Simulation of the distribution of solar heat in buildings using FormIt Pro software. MARKA (Media Arsitektur dan Kota): Jurnal Ilmiah Penelitian. 2024;**8**(1):11-22. DOI: 10.33510/marka.2024.8.1.11-22

[12] Purwanto LMF, Palilih L. Heat Transfer Calculations on the Wall with Psi-Therm Software. Research Paper. Soegijapranata Catholic University; 2020

[13] PROPER Best Practices and Innovation 2021. Ministry of Environment and Forestry of the Republic of Indonesia [Online]. pp. 2-5. Available from: https://proper.menlhk.go.id/propercms/uploads/magazine/docs/buku/magazineBest_Practice_2021.pdf

[14] Bergman TL et al. Introduction to Heat Transfer. 6th ed. Hoboken, NJ, USA: Wiley; 2011

[15] Çengel YA, Ghajar AJ. Heat and Mass Transfer: Fundamentals and Applications. 5th ed. New York, NY, USA: McGraw-Hill; 2014

[16] Bentoumi L et al. Impact of thermal insulation on energy consumption in buildings. Journal of Thermal Engineering. 2024;**10**(6):237-244. DOI: 10.14744/thermal.0000841

[17] Aljawad RH et al. Thermal performance analysis of local building materials for energy efficiency in Iraq. Journal of Thermal Engineering. 2024;**10**(4):1011-1020. DOI: 10.14744/thermal.0000845

[18] Nematchoua MK et al. Energy efficiency and thermal performance of office buildings integrated with passive strategies in coastal regions of humid and hot tropical climates in Madagascar. Applied Sciences. 2020;**10**(7). DOI: 10.3390/app10072438

[19] Latif E et al. Thermal Insulation Materials for Building Applications. London: ICE Publishing; 2019

[20] Granqvist CG. Handbook of Inorganic Electrochromic Materials. Amsterdam, Netherlands: Elsevier; 2002

[21] Hens T. Building Physics – Heat, Air and Moisture: Fundamentals and Engineering Methods with Examples and Exercises. 3rd ed. New York, NY, USA: Wiley; 2017

[22] Bhadani G. Building Materials for Sustainable Construction. Traverse City, Michigan, USA: Independently Published; 2021

[23] Lamond JF, Pielert J. Significance of Tests and Properties of Concrete and Concrete-Making Materials. 1st ed. West Conshohocken, PA, USA: ASTM International; 2006

[24] Bennett D. Sustainable Concrete Architecture. London, UK: Taylor & Francis; 2008

[25] Kavya Roofing Industries. Exploring the Versatility and Benefits of UPVC Roofing Sheets. Available from: https://www.kavyaroofing.com/exploring-the-versatility-and-benefits-of-upvc-roofing-sheets/

[26] Koumbem WND. Numerical study of the thermal performance of three roof models in hot and dry climates. Modeling and Numerical Simulation of Material Science. 2021;**11**:35-46. DOI: 10.4236/mnsms.2021.112003

[27] Viraat Industries. How UPVC Roofing Sheets Perform in Various Weather Conditions. Available from: https://viraatindustries.com/how-upvc-roofing-sheets-perform-in-various-weather-conditions/

[28] WC Online. All Things Gypsum: Using Gypsum Board as a Thermal Barrier. Available from: https://www.wconline.com/articles/85469-all-things-gypsum-using-gypsum-board-as-a-thermal-barrier

[29] Clarke JA. Energy Simulation in Building Design. 2nd ed. Routledge; 2001

[30] Hazal K. Benefits of Thermal Insulation in Roofs. Insulation Institute. Available from: https://sunparadise.com/benefits-of-thermal-insulation/

Chapter 8

Drying of Aqueous Films, an Application of Heat and Mass Transfer

Rodolphe Heyd, Julie Fichot, Driss Lahboub, Abderrahim Bakak, Christophe Josserand and Marie-Louise Saboungi

Abstract

Understanding the drying behavior of thin aqueous films is critical for a wide range of applications, from coatings and cosmetics to pharmaceutical and biotechnological processes. In this chapter, we develop a systemic framework to model the isothermal and isobaric drying of thin aqueous films in the absence of a convective boundary layer. Special emphasis is placed on the pivotal role of water activity, which directly governs the interfacial vapor pressure, evaporation kinetics, temporal evolution of film thickness, and final film composition. The study combines experimental observations, theoretical modeling, and numerical simulations to provide a detailed description of mass transport, interfacial dynamics, thermodynamic effects, and phase change phenomena involved in the drying process. The water-glycerol mixture is employed as a model system due to its well-characterized physicochemical properties and its relevance to a wide range of industrial formulations. The predictive capabilities of the model are validated against gravimetric measurements, which highlight the impact of solute enrichment on drying dynamics. This integrated framework offers a powerful predictive tool for the rational design of drying processes in multicomponent aqueous systems.

Keywords: drying, thin films, aqueous solutions, water chemical activity, advecto-diffusion equation, boundary immobilization method

1. Introduction

The drying of thin films composed of complex solutions is a ubiquitous phenomenon in a wide range of natural and industrial processes, spanning from coating technologies to pharmaceutical and cosmetic applications. It involves competing physical mechanisms—mass transfer, interfacial dynamics, thermodynamics, and phase change—whose interplay governs both the evolution of concentration profiles and the final structure of the film. Despite its technological importance, a detailed understanding of these mechanisms and accurate prediction of film behavior remain major challenges, due to the inherent complexity of multicomponent systems and associated nonideal effects.

Evaporation and solvent exchange profoundly influence the structuration of systems during drying. This effect is particularly pronounced in membranes, cosmetic films, and coatings, where drying can lead to the formation of blisters, perforations, or cracks [1–3]. Similarly, in cellular, plants, and animal biology as well as medicine, water transfer plays a critical role in maintaining cellular equilibrium [4].

While evaporation phenomena have been extensively studied from both theoretical and experimental perspectives, particularly for water droplets [5–7], less attention has been devoted to more complex systems such as colloidal suspensions, solutions, gels, and nonspherical geometries.

Two common drying regimes can be distinguished: one where mass transfer is governed by an external boundary layer due to forced convection over the film and another where no convective boundary layer is present. This latter case—despite its significant practical relevance—remains underexplored and is the focus of this study.

In thin-film configurations, mass diffusion, phase change dynamics, thermodynamics (through the chemical fugacity or activity of water), and viscoelastic relaxation phenomena (when present) coexist, interact, and compete to determine the drying kinetics and the final morphology of the dried film.

In this context, the development of predictive models capable of coherently describing the drying of complex solutions as thin films is a major scientific and technological challenge. A systemic approach, integrating the coupled effects of diffusion, evaporation, interfacial dynamics, and thermodynamics, is essential to elucidate the critical drying regimes and guide the rational design of controlled processes.

The water-glycerol mixture has been selected in this study as a representative model system, due to its well-characterized physicochemical properties, its industrial relevance, and the richness of its transport and evaporation behavior. Through an approach combining experimental observations, theoretical modeling, and numerical resolution, a detailed analysis of the mechanisms governing the isothermal and isobaric drying of this system without a convective boundary layer is proposed in this work.

Figure 1 presents a conceptual diagram summarizing the main physical phenomena involved during the drying of complex aqueous thin films. It highlights the competing processes—mass diffusion, thermodynamic effects (water chemical activity), viscoelastic relaxation, and phase change dynamics—and their collective role in controlling the interfacial dynamics, drying kinetics, and final properties of dried films.

This work thus aims to establish a systemic understanding of isothermal and isobaric drying of thin complex-solution films in the absence of a boundary layer. It also seeks to identify the key parameters governing evaporation kinetics and to lay the groundwork for the development of predictive models applicable to more elaborated multicomponent systems.

The modeling approach presented here can be applied, for example, to the isothermal and isobaric drying of:

- pure water thin films (see Section 3 and references [8, 9]);
- aqueous solution films modeled with water-glycerol mixtures (see Section 4);
- cosmetic thin films containing biocompatible polymers [10];
- biological thin films containing sugars [11–13].

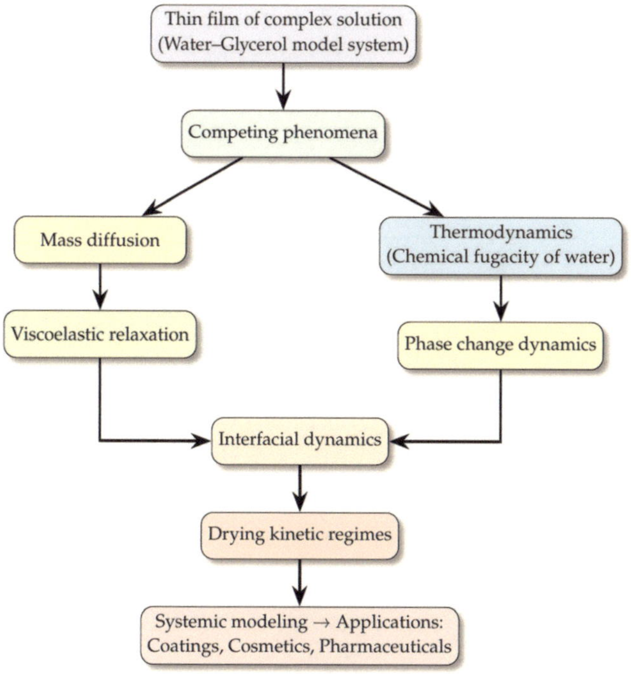

Figure 1.
Systemic modeling of the isothermal, isobaric, and boundary layer-free drying of complex aqueous thin films reveals that mass diffusion, water activity-driven thermodynamics, viscoelastic relaxation, and phase change dynamics collectively control interfacial evolution, drying kinetics, and the final film properties.

The remainder of the chapter is organized as follows. Section 2 presents the experimental setup developed for the controlled study of isothermal and isobaric drying of aqueous thin films. Section 3 introduces a model for the drying of pure water films and compares the modeling results with experimental data. Section 4 extends the model to aqueous solution films. Finally, the conclusions and perspectives are outlined.

2. Experimental approach

The drying dynamics of thin aqueous films occur over relatively short timescales (ranging from several minutes to a few hours) and are highly sensitive to ambient perturbations, such as temperature and humidity fluctuations. Therefore, a precise study of the drying process requires specific experimental precautions to ensure full environmental control throughout the experiment. It is generally not possible to accurately study the drying of aqueous films under ambient laboratory conditions or even inside conventional climatic chambers, as these are often ventilated in ways that are difficult to regulate.

2.1 Dedicated environmental chamber

The setup illustrated in **Figure 2** was specifically designed to enable the *in situ* preparation of thin aqueous films and the controlled study of their drying under

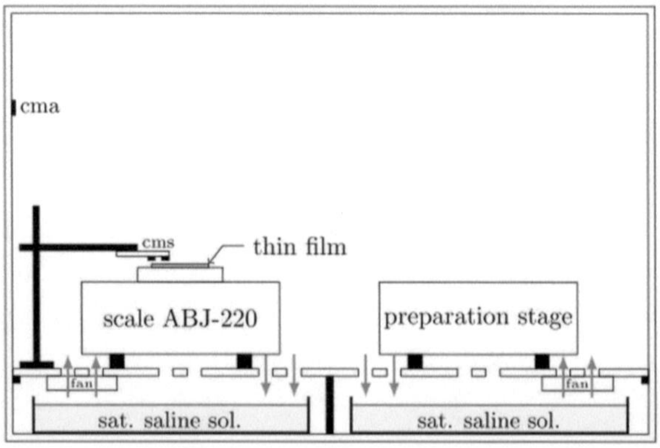

Figure 2.
Schematic of the environmental chamber designed for the controlled drying of thin aqueous films. Key components: cms = miniature surface environmental sensor; cma = miniature environmental sensor; sat. saline sol. = saturated saline solution. Access pathways to the internal volume are omitted for clarity.

well-defined temperature and humidity conditions while avoiding macroscopic air movement [14].

Brief description of the environmental chamber:

- The chamber, made from polymethyl methacrylate (PMMA) panels, can tightly enclose all the instrumentation necessary for the gravimetric study of thin film drying in a controlled atmosphere.

- A perforated floor supports the instruments (balance, preparation stage, sensors, and chemical solutions) while allowing for the regulation of the relative humidity inside the chamber *via* saturated saline solutions placed underneath. Circulating fans are used to accelerate vapor pressure homogenization within the chamber before experiments begin. The fans are switched off prior to measurements to prevent uncontrolled forced convection.

- An external heating/cooling system (not shown) enables regulation of the ambient temperature within the chamber. A simple radiative shield is employed to limit external thermal radiation and minimize greenhouse effects inside the enclosure.

- Miniature environmental sensors (cms) monitor the vapor partial pressure p_w^{int} and the temperature T near the film/air interface, while additional sensors (cma) monitor the environment far from the film. All measurements are managed remotely *via* sealed communication ports.

2.2 Embedded heating system

In some cases, it is necessary to precisely control the temperature of the film during drying. A flexible solution involves using thin-film resistive heaters attached directly to the substrate supporting the film. These heaters raise their own temperature through Joule heating, thereby controlling the temperature of the film.

However, regulating film temperature *via* heating elements can be particularly delicate when using high-precision balances. Any mechanical linkage between the measurement plate and external elements (such as the power supply or control electronics) introduces potential errors in mass measurements $m(t)$. Furthermore, using a radiative heating source (e.g., an infrared lamp) would be imprecise and still require instrumentation attached to the sample holder.

To overcome these issues, an autonomous embedded heating device powered by a miniature battery has been developed. This device is directly placed on the balance without any mechanical coupling and can control the film temperature very precisely. The design is shown in **Figure 3**.

Brief description of the embedded heating system:

- The film temperature is imposed using a thin resistive heating element (hf) glued onto an insulating substrate.
- A thin copper plate (cp) is interposed between the heating element and the film to homogenize the temperature.
- A miniature temperature sensor, attached to the top face of the film support, monitors the film surface temperature T^{sup}.
- This temperature signal is sent to a miniature controller (e.g., Minco type), which adjusts the Joule heating current $i_J(t)$ accordingly.
- The entire system (controller + battery + heater) weighs less than 180 g, making it compatible with standard analytical balances (dynamic range of 220 g).

2.3 Illustrative experimental results

Figure 4 presents typical drying curves obtained using the heating system of **Figure 3** and the environmental chamber of **Figure 2**, for substrate temperatures T^{sup} of 298 K, 303 K, and 308 K. As expected, the substrate temperature significantly affects the drying kinetics: the higher the temperature difference $T^{\mathrm{sup}} - T^{\infty}$ (where T^{∞} is the ambient temperature away from the film), the faster the drying process. This is mainly attributed to the increase of the saturation vapor density ρ_w^{sat} with temperature at the film interface.

The aqueous films corresponding to the curves in **Figure 4** were all prepared with similar initial compositions and geometrical characteristics (initial thickness $e_0 = 30\mu\mathrm{m}$) and were deposited on identical glass substrates under the same ambient conditions: $\mathrm{RH}^{\infty} = 43\%$ and $T^{\infty} = 298$ K, where RH^{∞} is the relative humidity of

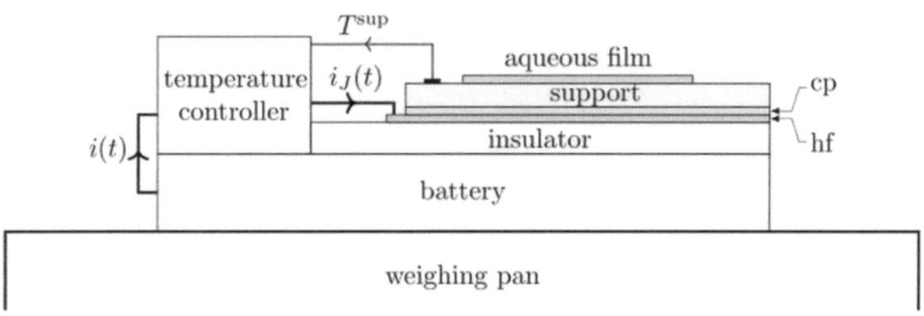

Figure 3.
Schematic of the heating device integrated onto the analytical balance platform, with: hf = heating film; cp = copper plate.

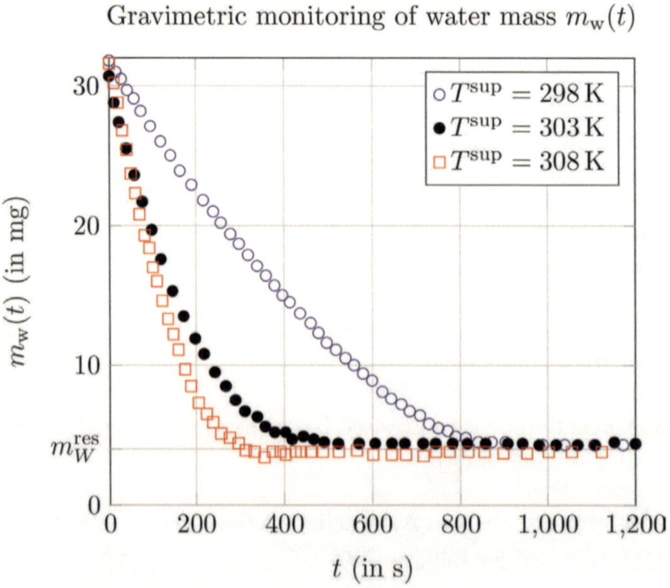

Figure 4.
Typical drying curves obtained for thin aqueous films (containing a water-soluble polymer) with very similar initial compositions and geometrical characteristics (thickness $e_0 = 30$ μm), under identical extreme conditions (ambient relative humidity $RH^\infty = 43\%$ and temperature $T^\infty = 298$ K), deposited onto identical substrates (glass slides). Only the substrate temperature T^{sup} differs between samples. A residual water mass m_W^{res} remains trapped within the polymer matrix at the end of the drying process.

ambient air far from the liquid film and T^∞ denotes the uniform temperature of the film as well as that of the surrounding environment, since the drying process is assumed to be isothermal.

Due to their small initial thickness, the drying dynamics are relatively fast (5–13 minutes), requiring continuous gravimetric monitoring. Notably, a residual water mass $m_W^{res} \neq 0$ remains trapped in the polymer matrix after drying, due to molecular interactions between water and the hydrophilic polymer (polyvinyl alcohol, PVA) used in these films.

3. Mathematical model for the drying of a pure water film

3.1 Problem formulation and assumptions

We consider the isothermal and isobaric evaporation of a planar liquid water film (denoted as W) in the absence of forced convection both within the film and at the liquid/gas interface with the surrounding humid air mixture (composed of water vapor (w) and dry air (da), see **Figure 5**). Both (w) and (da) are treated as ideal gases, and the dissolution of (da) into the liquid water is neglected.

The liquid film, characterized by a surface area Σ and an initial thickness $s_0 \equiv s(0)$, is assumed to be spread over a sufficiently wetting substrate, such that wetting effects can be neglected during most of the drying process. The film temperature T is maintained constant and equal to that of the surrounding atmosphere, for example, using the embedded heating device described in Section 2.2.

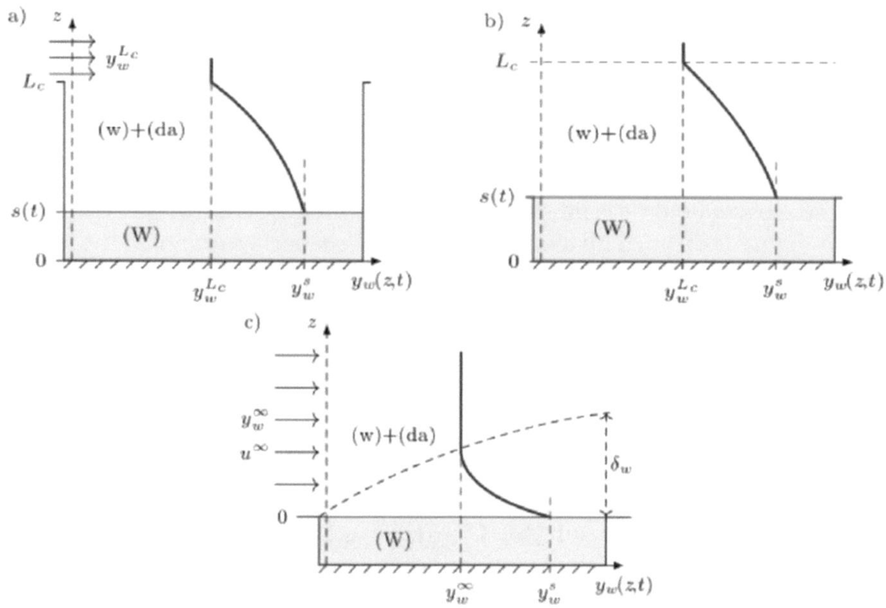

Figure 5.
Drying of a pure water film: (a, b) without forced convection; (c) with forced convection near the film. Temperature is regulated via an embedded heating device (see ***Figure** 3, not shown). δ_w: mass transfer boundary layer under forced convection.*

The case where evaporation occurs without any forced convection near the liquid/gas interface at $z = s(t)$ is the focus of the present mathematical modeling (see **Figure 5a** and **b**). In this configuration, the temporal evolution of the water mass $m_W(t)$ in the liquid film is governed solely by diffusion and free convection of water vapor in the stagnant humid air above the film. Neglecting edge effects, the mass transport within the humid air is considered one-dimensional, and the molar fractions y_w and y_{da} depend only on the vertical position z and time t.

3.2 Governing equations

A mass balance for the water vapor (w) species across an infinitesimal slab of thickness dz in the gas mixture leads to the following advection-diffusion equation (ADE):

$$\frac{\partial y_w}{\partial t} + v^*(t)\frac{\partial y_w}{\partial z} = D_0 \frac{\partial^2 y_w}{\partial z^2}, \quad \text{for} \quad z \in (s(t), L_c) \tag{1}$$

where:

- D_0 is the molecular diffusion coefficient of water vapor in air,
- L_c is the height of the drying chamber,
- $v^*(t)$ is the average molar velocity of the gas phase, given by:

$$v^*(t) = -D_0 \frac{1}{1 - y_w^s} \frac{\partial y_w}{\partial z}\bigg|_{z=s(t)} \tag{2}$$

Due to the assumptions of ideal gas behavior, constant pressure, and temperature, the total molar concentration $c = p/RT$ is constant.

Boundary and initial conditions:

- At $z = L_c$, the vapor molar fraction is set to $y_w(L_c, t) = y_w^{L_c}$,
- At the moving liquid/gas interface $z = s(t)$, the mole fraction of species (w) in the gas mixture is dictated by temperature and its chemical fugacity in the liquid film, thereby enforcing the boundary condition at the interface $y_w(s(t), t) = y_w^s(t)$,
- Initially, the vapor molar fraction is uniform: $y_w(z, 0) = y_w^{L_c}$ for $s(0) < z \leqslant L_c$.

In addition, a jump condition must be introduced to link the mass loss from the liquid film to the vapor flux at the moving boundary $z = s(t)$.

The total molar flux $\Phi_w^{*,t}$ of water leaving the liquid phase is given by:

$$\Phi_w^{*,t} = \mathbf{J}_w^{*,t}(s(t), t) \cdot \mathbf{e}_z \Sigma \tag{3}$$

where $\mathbf{J}_w^{*,t}$ is the total molar flux vector for water vapor at the interface, given by:

$$\mathbf{J}_w^{*,t}(s(t), t) = c y_w \mathbf{v}^*(t) - c D_0 \left.\frac{\partial y_w}{\partial z}\right|_{z=s(t)} \mathbf{e}_z \tag{4}$$

Using the conservation of mass and assuming no dissolution of dry air, the evolution of the interface position $s(t)$ is governed by:

$$\dot{s}(t) = -\frac{M_w c}{\rho_W^L} v^*(t) \tag{5}$$

where ρ_W^L is the mass density of liquid water and M_w is the molar mass of water. Substituting $\mathbf{v}^*(t)$, we obtain the jump condition:

$$\dot{s}(t) = \frac{M_w p D_0}{\rho_W^L R T} \frac{1}{1 - y_w^s} \left.\frac{\partial y_w}{\partial z}\right|_{z=s(t)} \tag{6}$$

3.3 Numerical solution using front immobilization

To facilitate the numerical resolution, the moving boundary problem is transformed *via* a change of variables:

$$x = \frac{z - s(t)}{L_c - s(t)} \tag{7}$$

mapping the physical domain $z \in (s(t), L_c)$ into a fixed domain $x \in (0, 1)$.

Defining $v_w(x, t) = y_w(x(L_c - s(t)) + s(t), t)$, the advection-diffusion equation becomes:

$$\frac{\partial v_w}{\partial t} = \frac{D_0}{(L_c - s(t))^2} \frac{\partial^2 v_w}{\partial x^2} + \frac{(1 - x)\dot{s}(t) - v^*(t)}{L_c - s(t)} \frac{\partial v_w}{\partial x} \tag{8}$$

with updated boundary and initial conditions:

$$v_w(0,t) = y_w^s(t), \quad v_w(1,t) = y_w^{L_c}, \quad v_w(x,0) = y_w^{L_c} \tag{9}$$

and jump condition:

$$\dot{s}(t) = \frac{M_w p D_0}{\rho_w^L RT} \frac{1}{(1 - y_w^s)(L_c - s(t))} \frac{\partial v_w}{\partial x}\bigg|_{x=0} \tag{10}$$

A fully implicit finite-difference scheme on a uniform grid x_i is used to solve the system numerically (see Appendix A). The discrete problem at each time step can be written in matrix form, and the evolution of $s(t)$ is computed accordingly.

3.4 Common approximations

In the modeling of evaporation-driven mass transport in humid air, several classical approximations are frequently used to simplify the advection-diffusion problem. These approximations are evaluated below in the context of thin-film drying.

3.4.1 Stationary approximation (SA)

The stationary approximation assumes that:

- The vaporization front remains fixed at $s(t) = s_0$,
- The advection-diffusion equation has reached a stationary regime, meaning time derivatives can be neglected.

Under these conditions, an analytical expression for the water vapor molar fraction $y_w^{\mathrm{SA}}(z)$ in the gas phase can be derived:

$$y_w^{\mathrm{SA}}(z) = 1 - \left(1 - y_w^{s_0}\right) \left(\frac{1 - y_w^{L_c}}{1 - y_w^{s_0}}\right)^{\frac{z - s_0}{L_c - s_0}} \tag{11}$$

The corresponding time evolution of the film thickness is linear:

$$s^{\mathrm{SA}}(t) = s_0 + p_e t \tag{12}$$

where the drying rate p_e is:

$$p_e = -\frac{p M_w D_0}{RT \rho_W^L (L_c - s_0)} \ln\left(\frac{1 - y_w^{L_c}}{1 - y_w^{s_0}}\right) \tag{13}$$

3.4.2 Quasi-stationary approximation (QSA)

The quasi-stationary approximation relaxes the fixed-front assumption:

- The vaporization front $s(t)$ is allowed to move with time,
- The advection-diffusion equation has reached a stationary regime, meaning time derivatives can be neglected.

In this case, the vapor molar fraction and the film thickness evolve according to:

$$y_w^{\mathrm{QSA}}(z) = 1 - \left(1 - y_w^s\right)\left(\frac{1 - y_w^{L_c}}{1 - y_w^s}\right)^{\frac{z - s^{\mathrm{QSA}}(t)}{L_c - s^{\mathrm{QSA}}(t)}} \quad \text{and}$$
$$s^{\mathrm{QSA}}(t) = s_0 - (L_c - s_0)\left(\sqrt{1 - \frac{2p_e t}{L_c - s_0}} - 1\right) \tag{14}$$

This approximation captures the moving interface more accurately and will be shown to perform much better than SA, especially over extended drying periods.

3.4.3 Instationary approximation for an infinite cell (ICA)

This instationary approximation assumes:

- The drying chamber height is infinite $L_c \to \infty$,
- The front is stationary $s(t) = s_0$,
- The full time-dependence of the advection-diffusion equation is: retained.

The water vapor molar fraction is then given analytically by (see [15], Chapter 20):

$$y_w^{L_c,\infty}(z,t) = \frac{y_w^{s_0}}{1 + \mathrm{erf}(\xi)}\left[1 + \frac{y_w^{\infty}}{y_w^{s_0}}\mathrm{erf}(\xi) - \left(1 - \frac{y_w^{\infty}}{y_w^{s_0}}\right)\mathrm{erf}\left(\frac{z - s_0}{\sqrt{4D_0 t}} - \xi\right)\right] \tag{15}$$

where ξ is a solution of the following nonlinear algebraic equation:

$$\xi = \frac{1}{\sqrt{\pi}}\frac{y_w^{s_0}}{1 - y_w^{s_0}}\frac{1 - y_w^{\infty}/y_w^{s_0}}{1 + \mathrm{erf}(\xi)}e^{\xi^2} \tag{16}$$

This approach can be useful to describe the very initial stages of drying (short times) but becomes increasingly inaccurate as time progresses, because it neglects both front movement and chamber size effects, making it unsuitable for practical thin-film configurations.

In the following sections, the results from the full numerical model will be compared to those obtained using these approximations, in order to quantify their validity domains.

3.5 Applications and main results

3.5.1 Materials and methods

The mathematical model presented earlier is applied to the isothermal ($T = 298.15$ K) and isobaric ($p = 1$ atm) drying of thin pure water films. The liquid films are prepared using ultrapure water (MilliQ, Millipore Corporation, Bedford, MA) and deposited on filter paper substrates (Whatman grade 93, pore size 10 μm) with an exchange area $\Sigma = 56$ mm^2. The drying chamber height is set to $L_c = 3.45$ mm.

Experimental mass loss measurements are performed gravimetrically using a high-precision analytical balance (ABJ 220-4NM, Kern). The experimental results are then compared to predictions from:

- The full numerical model (referred to as Model),
- The stationary approximation (SA) and the quasi-stationary approximation (QSA).

Boundary conditions:

- The molar fractions y_w^s and $y_w^{L_c}$ are deduced from the water activity a_w measured using a precision miniature thermo-hygrometric sensor (SHT85, Sensirion).
- Water activity relates to vapor pressure through:

$$a_w = \frac{p_w}{p_w^{\text{sat}}(T)} = y_w \frac{p}{p_w^{\text{sat}}(T)} \tag{17}$$

where $p_w^{\text{sat}}(T)$ is the saturation vapor pressure of pure water at temperature T, retrieved from ASHRAE standard tables [16].

Water vapor diffusion coefficient: The diffusion coefficient D_0 is evaluated using the Marrero-Mason correlation [17]:

$$D_0(T, p) = 1.87 \times 10^{-10} T^{2.072} \left(\frac{p_{\text{ref}}}{p}\right) \tag{18}$$

with T the temperature in K, p the pressure in Pa, and $p_{\text{ref}} = 101325$ Pa.

3.5.2 Temporal behavior

Thanks to the numerical model, it is possible to examine the behavior of the system at extremely short timescales—much shorter than the characteristic diffusion time $\tau_0 = s_0^2/D_0$ — which are inaccessible experimentally due to gravimetric limitations.

Figure 6 illustrates two extreme temporal regimes observed for an initial thickness $s_0 = 539 \mu\text{m}$ (with $\tau_0 \approx 12$ ms) and a total drying duration of about 52 minutes.

Short-time behavior ($t \ll \tau_0$).

At very early times (see **Figure 6a**):

- The full numerical model (Model) and the instationary approximation (ICA) are in excellent agreement.
- However, both the stationary (SA) and quasi-stationary (QSA) approximations fail to correctly describe the vapor concentration profile.

Thus, SA and QSA cannot be used to model the initial transient phase.

Long-time behavior ($t \gg \tau_0$).

At longer times (see **Figure 6b**):

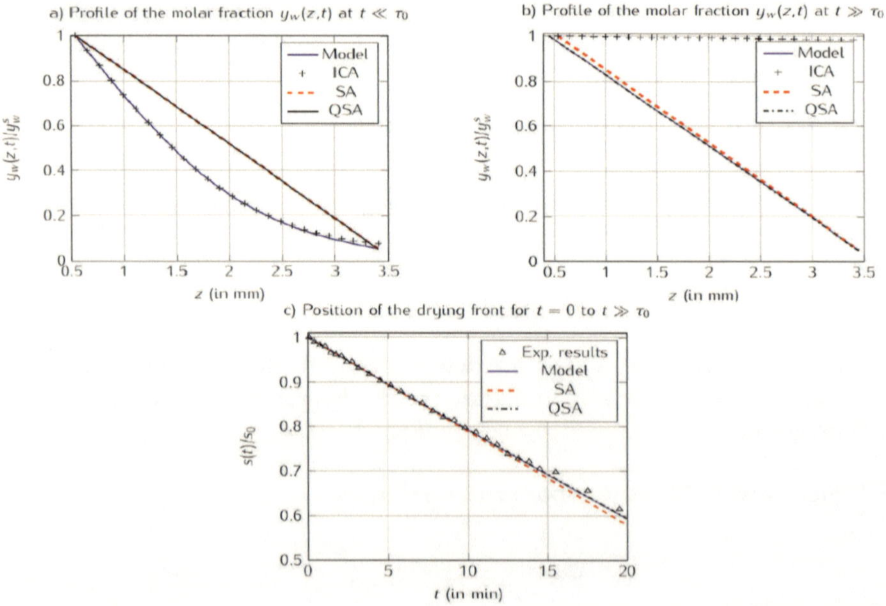

Figure 6.
Temporal evolution during the drying of a pure water film. (a and b) Water mole fraction profiles $y_w(z,t)$; (c) Evaporation front position (or liquid film thickness). The physical parameters (L_c, D_0, y_w^s, $y_w^{L_c}$, p, T) are identical for the three approaches (Model, SA, and QSA).

- The ICA becomes inaccurate because it neglects the moving front and the finite size of the chamber.

- In contrast, both the Model and the QSA approximation correctly capture the quasi-linear vapor profile across the gas phase.

Moreover, the vapor molar fraction profile becomes almost perfectly linear during most of the drying process. This allows for the use of coarser spatial discretizations in the numerical solution, substantially reducing computational cost.

A useful approximate expression for the vapor molar fraction at long times can be deduced from the preceding discussion:

$$y_w(z,t) = \frac{y_w^{L_c} - y_w^s}{L_c - s(t)}(z - s(t)) + y_w^s \tag{19}$$

The evolution of the evaporation front $s(t)$ is a key feature for validating the model. **Figure 6c** shows that:

- the stationary approximation (SA) significantly underestimates the drying time, due to neglecting the progressive increase of $L_c - s(t)$,

- the Model and the QSA approximation both provide an excellent match with experimental measurements across the entire drying period.

Consequently, in the remainder of the study, only the full Model and the QSA will be used to describe thin-film drying under the present conditions.

3.5.3 Drying of thin pure water films

The experimental validation of the model is performed by comparing gravimetric drying data of pure water films to the model predictions, under strictly controlled isothermal ($T = 298.15$ K) and isobaric ($p = 1$ atm) conditions, without any convective boundary layer.

3.5.3.1 Experimental results

Figure 7 presents the drying curves for two pure water films deposited on filter paper supports. Both films have comparable geometrical characteristics: same surface area $\Sigma = 56$ mm^2; same chamber height $L_c = 3.45$ mm; very similar initial thicknesses $s_0 = 539$ μm and $s_0 = 514$ μm.

3.5.3.2 Analysis and discussion

3.5.3.2.1 Drying kinetics

Figure 7a and **b** show the evolution of the film thickness $s(t)$ over time. Key observations are as follows:

- The water activity $a_w^{L_c}$ of the ambient vapor phase is confirmed as a critical parameter governing the drying kinetics in the absence of a boundary layer,
- Only the Model and QSA approaches accurately predict the drying dynamics throughout the entire experimental time domain,
- As expected, the film exposed to the lower ambient humidity (lower $a_w^{L_c}$) dries significantly faster.

3.5.3.2.2 Quality of the approximations

The stationary approximation (SA) consistently underestimates the total drying time. This is because its drying rate p_e, derived from a fixed front assumption, does

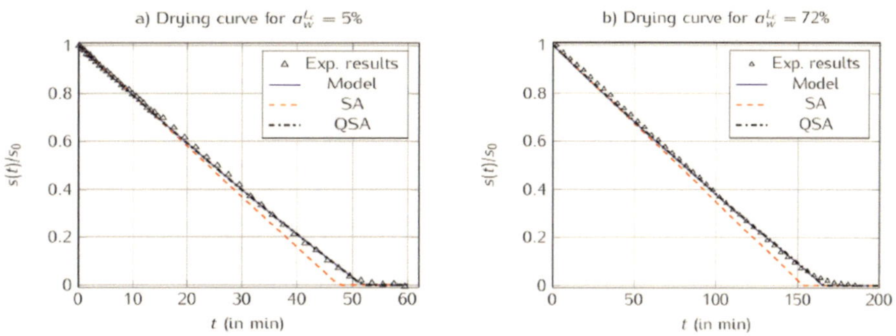

Figure 7.
Isothermal ($T = 298.15$ K) and isobaric ($p = 1$ atm) drying of pure water films ($\Sigma = 56$ mm^2, $L_c = 3.45$ mm): (a) $s_0 = 539$ μm, $a_w^{L_c} = 0.05$; (b) $s_0 = 514$ μm, $a_w^{L_c} = 0.72$.

not account for the progressive increase of the gas phase thickness $(L_c - s(t))$, which progressively slows down the evaporation rate.

The QSA correctly incorporates this effect and therefore provides an excellent match to experimental observations.

3.5.3.2.3 Influence of the substrate

At late stages of drying (approximately 52 min for $a_w^{L_c} = 0.05$ and 160 min for $a_w^{L_c} = 0.72$), a slight slowing down of the evaporation rate is observed. This effect is attributed to interactions between the porous filter substrate and the remaining liquid, not accounted for in the current model or in the QSA approximation.

3.5.3.2.4 Mass transfer coefficients

Even though no convective mass boundary layer is present, it is still possible to define an instantaneous mass transfer coefficient $h_m^X(t)$ relative to any transported quantity X, using the relation $\Phi_m^X(t) = h_m^X(t)\Sigma(X_w^s - X_w^{L_c})$. In the case of the vapor molar fraction y_w, the analytical expression becomes:

$$h_m^{y_w}(t) = \frac{M_w D_0 p}{RT} \frac{1}{(L_c - s(t))(1 - y_w^s)} \tag{20}$$

This result will be particularly useful for modeling the drying of aqueous solutions in the next section.

3.5.3.2.5 Simplified linear model

As highlighted earlier, the vapor molar fraction profile in the gas phase becomes quasi-linear for most of the drying period $\tau_0 \ll t \leqslant t_s$. Thus, for practical purposes, the following simplified expression for $y_w(z,t)$ is highly accurate:

$$y_w(z,t) \approx \frac{y_w^{L_e} - y_w^s}{L_c - s(t)}(z - s(t)) + y_w^s \tag{21}$$

This approximation enables efficient numerical resolution with low computational cost, without compromising accuracy. **Figure 8** compiles these results and displays the spatial profile of the water mole fraction arising from the isothermal and isobaric drying of a pure water film.

4. Extension to the drying of aqueous solutions

The model previously developed is now extended to the drying of an aqueous film containing a fully miscible and very low-volatility solute, such as glycerol (or, e.g., a sugar or a polymer), whose vapor fraction can be neglected in the gas phase above the film.

4.1 Generalization of the model

We still consider isothermal and isobaric drying without a convective boundary layer, but now, the liquid phase contains a solute, glycerol (Gl), perfectly miscible

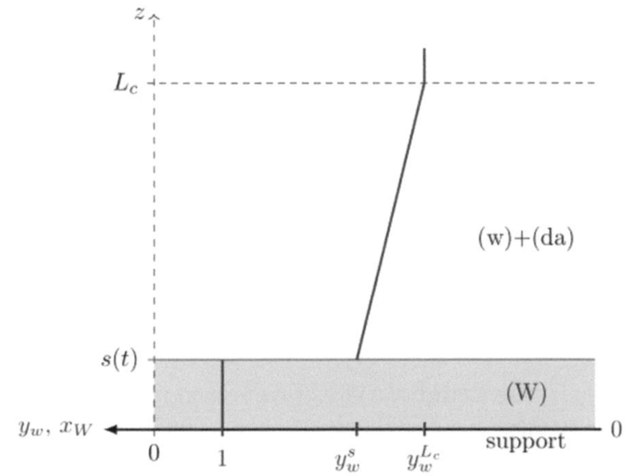

Figure 8.
Water molar fraction profiles (y_w and x_W, blue lines) for a thin pure water film (W) undergoing isothermal and isobaric drying without a convective boundary layer, at an intermediate time satisfying $\tau_0 \ll t < t_s$. Spatial and molar fraction scales are not preserved for clarity of graphical representation.

with water (W) at all proportions. Glycerol is widely used across industries—from food processing to chemical manufacturing and heat transfer applications—where it provides an environmentally friendly, biodegradable alternative to ethylene glycol. The well-documented physical properties of both water and glycerol make the water-glycerol mixture an ideal model system for studying the drying of binary liquid films.

The liquid mixture forming the thin film is now denoted (W) + (Gl). Given the extremely low volatility of glycerol compared to water, the vapor phase above the film consists exclusively of water vapor (w) and dry air (as), and is denoted (w) + (as).

As drying progresses, the molar concentration x_{Gl} of glycerol within the liquid phase increases, giving rise to mass transport phenomena within the liquid film. Following Hennessy *et al.* [18], we assume that Fickian diffusion is the sole mode of mass transport within the liquid phase. Moreover, given the relatively close values of the densities of the solvent ($\rho_W^L = 997\ \mathrm{kg/m^3}$, from [16]) and the solute ($\rho_{\mathrm{Gl}} = 1258\ \mathrm{kg/m^3}$, from reference [19]), we neglect variations in the liquid mixture density ρ_L, assuming it remains constant throughout drying. Volume effects are also neglected.

Since the liquid composition evolves during drying, the chemical fugacity f of water changes accordingly and must be computed at each instant to evaluate the evaporative flux at $z = s(t)$. Local thermodynamic equilibrium is assumed at the moving interface $z = s(t)$ between water (W) in the liquid phase and water vapor (w) in the gas phase, expressed as:

$$f_W(T,p) = f_w(T,p), \quad \text{at} \ \ z = s(t) \tag{22}$$

Assuming ideal behavior for gases (w) and (as), the gas mixture (w) + (as) follows the Lewis-Randall rule, allowing $f_w(T,p) = p_w = y_w p$, where y_w is the mole fraction of water vapor in the gas phase.

For the liquid mixture, the water fugacity $f_W(T,p)$ is expressed as:

$$f_W(T,p) = x_W \gamma_W(T,p) f_W^*(T,p) \tag{23}$$

where x_W and γ_W are the mole fraction and activity coefficient of water in the liquid phase (W) + (Gl), respectively. The fugacity f_W^* of pure water at temperature T and pressure p is equal, under typical drying conditions, to the saturated vapor pressure $p_w^{sat}(T)$:

$$f_W^*(T,p) = p_w^{sat}(T) \tag{24}$$

Thus, at the interface $z = s(t)$:

$$y_w = x_W \gamma_W \frac{p_w^{sat}}{p} \tag{25}$$

The water activity a_W in the liquid can thus be defined as $a_W = x_W \gamma_W$, and recalling that the vapor phase activity is $a_w = \frac{p_w}{p_w^{sat}}$, we deduce:

$$a_W = a_w \quad \text{or equivalently} \quad p_w = a_W p_w^{sat} \tag{26}$$

Thus, at constant drying temperature, the partial pressure p_w of water vapor—and hence the volatility of water—at the interface decreases as the water activity a_W in the liquid diminishes. Consequently, a_W at the interface becomes a key parameter governing the drying kinetics of aqueous solutions.

4.2 Modeling the chemical activity of water

The chemical activity a_i of a component (i) in a liquid mixture characterizes the intensity of chemical interactions between the molecules of (i) and those of the other components present. For a pure liquid, the reference activity a_i^* is set to unity, $a_i^* = 1$. The presence of solutes $(j) \neq (i)$ generally leads to a decrease in the chemical activity a_i of (i) within the mixture: $0 \leq a_i \leq a_i^* = 1$.

The chemical activity a_i of a constituent in a liquid mixture mainly depends on the temperature T, the nature of the components, and the composition $\{x_i\}$, where x_i is the mole fraction of (i): $a_i = a_i(T, \{x_i\})$.

4.2.1 Types of chemical activity models

There are primarily four approaches to determine $a_i(T, \{x_i\})$: experimental measurements, descriptive modeling (based on empirical data), predictive modeling (based on theoretical considerations), and molecular simulations (e.g., molecular dynamics).

Descriptive or predictive approaches typically involve expressing the molar excess Gibbs free energy g^E of the liquid mixture as a function of temperature and composition: $g^E = g^E(T, \{x_i\})$.

Among the commonly cited approaches (see, e.g., [20, 21]), two main categories are generally distinguished: descriptive approaches and predictive approaches. The following lists present representative examples of each category.

- Descriptive: Van Laar, Margules, Flory-Huggins lattice models, Non-Random Two-Liquids (NRTL) [22], and UNIversal QUAsi-Chemical (UNIQUAC) [23],

- Predictive: UNIQUAC Functional-group Activity Coefficients (UNIFAC) [24] and Modified UNIFAC [25].

In this study, the chemical activity of water in the (W) + (Gl) system is modeled descriptively using the NRTL approach, based on experimental data reported by Zaoui et al. [26].

4.2.2 Predictive NRTL modeling

The Non-random Two-Liquids (NRTL) model was originally proposed by Prausnitz and Renon in 1968 [22]. It is a descriptive model characterized by several parameters (initially three for a binary mixture), whose values are determined by fitting experimental data for a given liquid system. The molar excess Gibbs-free energy g^E_{NRTL} is expressed, for a mixture of M components, as (see [20]):

$$g^E_{\mathrm{NRTL}} = RT\sum_{i=1}^{M} x_i \frac{\sum_j \tau_{ji} G_{ji} x_j}{\sum_j G_{ji} x_j} \tag{27}$$

with:

$$\left| \begin{array}{l} G_{ij} = \exp(-\alpha_{ij}\tau_{ij}) \;\text{ and }\; G_{ii} = 1 \\ \tau_{ij} = A_{ij} + \dfrac{B_{ij}}{T} \;\; ; \;\; \alpha_{ij} = \alpha_{ji} \;\; ; \;\; \tau_{ii} = 0 \end{array} \right. \tag{28}$$

where A_{ij}, B_{ij}, and α_{ij} are adjustable interaction parameters.

The activity coefficient γ_i of component (i) is related to g^E_{NRTL} *via*:

$$\ln \gamma_i = \frac{1}{RT}\left(\frac{\partial g^E_{\mathrm{NRTL}}}{\partial x_i}\right)_{T,p,x_{j\neq i}} \tag{29}$$

From the NRTL model, the explicit expression of $\ln \gamma_i$ becomes:

$$\ln \gamma_i = \frac{\sum_{j=1}^{M} x_j \tau_{ji} G_{ji}}{\sum_{j=1}^{M} x_j G_{ji}} + \sum_{j=1}^{M} \frac{x_j G_{ij}}{\sum_{k=1}^{M} x_k G_{kj}} \left(\tau_{ij} - \frac{\sum_{m=1}^{M} x_m \tau_{mj} G_{mj}}{\sum_{k=1}^{M} x_k G_{kj}}\right) \tag{30}$$

For a binary system $(W) + (Gl)$ $(M = 2)$, Eqs. (27) and (30) simplify as:

$$\left| \begin{array}{l} g^E_{\mathrm{NRTL}} = x_1 x_2 RT\left(\dfrac{\tau_{21}G_{21}}{x_1 + x_2 G_{21}} + \dfrac{\tau_{12}G_{12}}{x_2 + x_1 G_{12}}\right) \\ \ln \gamma_1 = x_2^2\left[\tau_{21}\left(\frac{G_{21}}{x_1 + x_2 G_{21}}\right)^2 + \dfrac{\tau_{12}G_{12}}{(x_2 + x_1 G_{12})^2}\right] \\ \ln \gamma_2 = x_1^2\left[\tau_{12}\left(\frac{G_{12}}{x_2 + x_1 G_{12}}\right)^2 + \dfrac{\tau_{21}G_{21}}{(x_1 + x_2 G_{21})^2}\right] \end{array} \right. \tag{31}$$

where component (1) corresponds to water (W) and component (2) to glycerol (Gl). Thus, for the binary system (W) + (Gl), five NRTL parameters must be determined from experimental data: A_{WGl}, A_{GlW}, B_{WGl}, B_{GlW}, and $\alpha_{\mathrm{WGl}} = \alpha_{\mathrm{GlW}}$.

Within the framework of the NRTL model, the water activity $a_W = x_W \gamma_W$ in the binary mixture (W) + (Gl) is given by:

$$a_W = x_W \exp\left\{(1-x_W)^2\left[\tau_{\mathrm{GlW}}\left(\frac{G_{\mathrm{GlW}}}{x_W+(1-x_W)G_{\mathrm{GlW}}}\right)^2+\frac{\tau_{\mathrm{WGl}}G_{\mathrm{WGl}}}{((1-x_W)+x_W G_{\mathrm{WGl}})^2}\right]\right\} \tag{32}$$

where the parameters A_{WGl}, A_{GlW}, B_{WGl}, B_{GlW}, and α_{WGl} are obtained from the work of Song et al. [27] based on experimental data from Zaoui et al. [26].

Figure 9 shows the variation of the water activity a_W as a function of the water mole fraction x_W for different temperatures. It is observed that while a_W varies significantly with x_W, it remains relatively insensitive to temperature over the range $T \in [273.15, 353.15]$ K.

4.3 Setting up the equations

The key elements of the modeling considered in this section are schematically illustrated in **Figure 10**.

4.3.1 Assumptions and governing equations

The study of isothermal and isobaric drying of a pure water film (presented in Section 3) showed that the molar fraction y_w of water vapor in the gas phase above the film can be approximated by a quasi-stationary linear profile during the drying period $\tau_0 \ll t < t_s$, as shown in **Figure 10**:

$$y_w(z,t) = \frac{y_w^{L_c} - y_w^s(t)}{L_c - s(t)}(z - s(t)) + y_w^s(t), \quad \text{for } z \in (s(t), L_c] \text{ and } \tau_0 \ll t < t_s \tag{33}$$

where τ_0 is typically less than one second, and t_s represents the total drying time, often ranging from several minutes to several hours. Therefore, it will now be assumed that $\tau_0 = 0$.

Assuming that the liquid mixture density ρ_L remains constant, it is more convenient to describe mass transport within the film using volume fractions ϕ_W and ϕ_{Gl} of water and glycerol, respectively, rather than molar fractions.

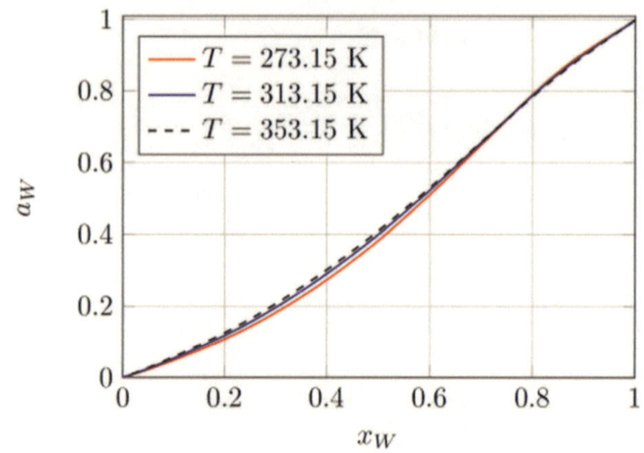

Figure 9.
Evolution of the water activity a_W in the (W) + (Gl) mixture as a function of the water mole fraction x_W for different temperatures.

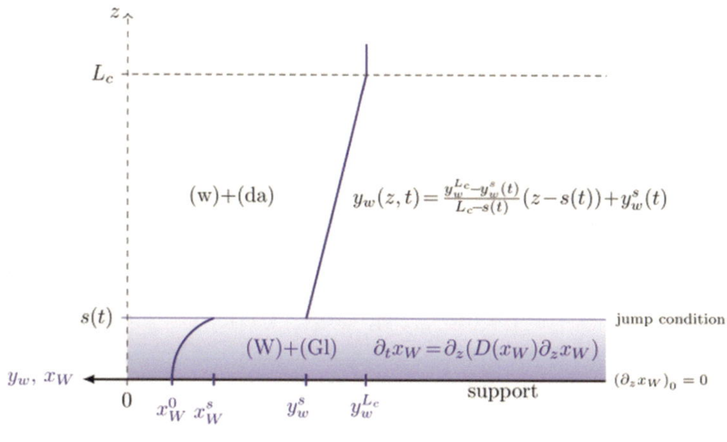

Figure 10.
Schematic representation of the molar fraction profiles of water x_W and vapor y_w (blue lines) for a thin aqueous film (W) + (Gl) subjected to isothermal and isobaric drying, without boundary layer, at a time t satisfying $\tau_0 \ll t < t_s$. Spatial and molar fraction scales are not respected for clarity.

Under the assumption of purely diffusive transport within the liquid film, the evolution of the water volume fraction ϕ_W is governed by the nonlinear diffusion equation, as shown in **Figure 10**:

$$\frac{\partial \phi_W}{\partial t} = \frac{\partial}{\partial z}\left(D(\phi_W)\frac{\partial \phi_W}{\partial z}\right), \quad \text{for } z \in (0, s(t)) \text{ and } 0 < t < t_s \tag{34}$$

where $D(\phi_W) = D_W^0 \chi(\phi_W)$ is the mutual diffusion coefficient, with $D_W^0 = 10^{-9}\,\text{m}^2/\text{s}$ and $\chi(\phi_W)$ a function of ϕ_W, adjusted from the experimental data of D'Errico *et al.* [28] at $T = 298$ K:

$$\chi(\phi_W) = \frac{0.015 + 0.528\phi_W}{1.11 - 0.58\phi_W} \tag{35}$$

The diffusion Eq. (34) is accompanied by an initial condition and two boundary conditions:

- *Initial condition*: At $t = 0$, the film is assumed homogeneous, with a uniform initial water volume fraction:

$$\phi_W(z, 0) = \phi_W^0 = \frac{1}{1 + V_{Gl}^0 / V_W^0} \tag{36}$$

where V_W^0 and V_{Gl}^0 are the initial volumes of water and glycerol, respectively.

- *Boundary condition at the substrate* ($z = 0$): due to the impermeable nature of the substrate, a no-flux Neumann condition applies:

$$\left.\frac{\partial \phi_W}{\partial z}\right|_{z=0} = 0 \tag{37}$$

A boundary condition must also be formulated at the moving interface $z = s(t)$, together with an evolution equation for the interface position.

A local mass balance for water at $z = s(t)$ gives the following jump condition:

$$\dot{m}_W(s) = \Phi_W^{d,m}(s) - \Phi_w^{t,m}(s) \tag{38}$$

with: $\dot{m}_W(s) = \Sigma\dot{s}\rho_W^L\phi_W$, where Σ is the liquid/gas interface area; $\Phi_W^{d,m} > 0$ is the diffusive mass flux of water within the liquid; and $\Phi_w^{t,m} > 0$ is the total advective-diffusive mass flux of water vapor in the gas phase.

Using Fick's first law and assuming negligible variation in the liquid density:

$$\Phi_W^{d,m}(s) = -D\rho_W^L\Sigma\left.\frac{\partial\phi_W}{\partial z}\right|_{s(t)} \tag{39}$$

The vapor flux $\Phi_w^{t,m}$ can be expressed as $\Phi_w^{t,m} = \rho_W^L J\Sigma$, where J is the evaporation rate at the interface $z = s(t)$. Thus, the jump condition at $z = s(t)$ reads:

$$-D(\phi_W)\left.\frac{\partial\phi_W}{\partial z}\right|_{s(t)} = J + \dot{s}\phi_W(s) \tag{40}$$

Different expressions for the evaporation rate J have been proposed in the literature. Okuzono et al. [29] adopted a constant value $J = J_0$ in the absence of skin formation at the interface $z = s(t)$, whereas Hennessy et al. [18] proposed a formulation of the form $J = k(\phi_W - \phi_W^e)$, where ϕ_W^e is the water volume fraction reached by the liquid film at equilibrium with the surrounding gas atmosphere, and k is a constant mass transfer coefficient. In the approach of Hennessy et al. [18], the parameters k and ϕ_W^e are adjusted based on experimental data.

In the present approach, the evaporative rate J is expressed based on the vapor mass transport modeling above the film exposed in Section 3:

$$J = \frac{pD_0M_W}{\rho_W^L RT}\frac{y_w^s - y_w^{L_c}}{(L_c - s(t))(1 - y_w^s)} = \frac{h_m^{y_w}}{\rho_W^L}\left(y_w^s - y_w^{L_c}\right) \tag{41}$$

where D_0 is the binary diffusion coefficient of water vapor in air, M_W the molar mass of water, p the total pressure, and $h_m^{y_w}$ the mass transfer coefficient. At the interface, y_w^s is computed from the local equilibrium assumption:

$$y_w^s = x_W^s\gamma_W(x_W^s)\frac{p_w^{\text{sat}}}{p} \tag{42}$$

where $\gamma_W(x_W^s)$ is evaluated using Eq. (32), and x_W^s is deduced from the volume fraction ϕ_W^s *via*:

$$x_W = \frac{M_{\text{Gl}}\rho_W^L\phi_W}{M_{\text{Gl}}\rho_W^L\phi_W + M_W\rho_{\text{Gl}}^L(1-\phi_W)} \tag{43}$$

Note that J depends both on the interface composition ϕ_W^s and the instantaneous film thickness $s(t)$. It should also be pointed out that, within the present modeling framework, there is strictly no parameter requiring experimental adjustment for the case shown in **Figure 5a**, and only one parameter—the domain length L_c—for the case in **Figure 5b**. This represents a major advantage of the present approach compared to that of Hennessy et al. [18].

A mass balance for glycerol at $z = s(t)$ leads to a second jump condition. Given the negligible volatility of glycerol:

$$\dot{s}(t)\phi_{\mathrm{Gl}} = -D\frac{\partial \phi_{\mathrm{Gl}}}{\partial z}\bigg|_{s(t)} \tag{44}$$

Using the binary nature of the mixture, this gives a second conservation relation for ϕ_W:

$$-D(\phi_W)\frac{\partial \phi_W}{\partial z}\bigg|_{s(t)} = -\dot{s}(t)(1-\phi_W) \tag{45}$$

Subtracting Eq. (40) from Eq. (45) yields the evolution equation for the interface:

$$\dot{s}(t) = -J \tag{46}$$

which is nothing else than the overall water balance for the aqueous film.

It is also necessary to examine the global mass balance for glycerol, which is conserved within the film at all times: $m_{\mathrm{Gl}}(t) = m_{\mathrm{Gl}}(0)$, or equivalently:

$$\rho_{\mathrm{Gl}}^L \Sigma \int_0^{s(t)} \phi_{\mathrm{Gl}}(z,t)\mathrm{d}z = \rho_{\mathrm{Gl}}^L \Sigma \phi_{\mathrm{Gl}}^0 s_0 \tag{47}$$

Recalling that the liquid mixture is binary, this global conservation of glycerol can be rewritten in the form of the following integral equation:

$$s(t) = \left(1-\phi_W^0\right)s_0 + \int_0^{s(t)} \phi_W(z,t)\mathrm{d}z \tag{48}$$

where $s_0 = s(0)$ and $\phi_W^0 = \phi_W(z,0) = \mathrm{cste}$. At equilibrium (end of drying), it is found from Eq. (48) that the equilibrium thickness s_e satisfies:

$$\frac{s_e}{s_0} = \frac{1-\phi_W^0}{1-\phi_W^e} \tag{49}$$

which provides a useful consistency check for numerical solutions.

4.3.2 Synthesis of the equations and immobilization of the moving front

Due to the highly nonlinear nature of the problem, no exact analytical solution is known to describe the complete drying dynamics across the entire time range. Consequently, a numerical resolution is necessary.

To simplify solving the problem, the moving front is immobilized by introducing the change of variable $x = z/s(t)$, with $x \in (0,1)$.

Applying this transformation to the diffusion Eq. (34) and using the chain rule, the governing equations become:

$$s^2(t)\frac{\partial \phi_W}{\partial t} = D_W^0 \frac{\partial}{\partial x}\left(\chi(\phi_W)\frac{\partial \phi_W}{\partial x}\right) + xs(t)\dot{s}(t)\frac{\partial \phi_W}{\partial x} \quad \text{for } x \in (0,1) \text{ and } 0 < t < t_s$$
$$\left.\frac{\partial \phi_W}{\partial x}\right|_{x=0} = 0 \text{ and } D_W^0 \chi(\phi_W)\left.\frac{\partial \phi_W}{\partial x}\right|_{x=1} = s(t)\dot{s}(t)[1-\phi_W(1,t)]$$
$$\dot{s}(t) = -J\left(\phi_W^s(t), s(t)\right) \text{ and } \chi(\phi_W) = \frac{0.015 + 0.528\phi_W}{1.11 - 0.58\phi_W} \tag{50}$$

A finite element discretization of type $\mathbb{P}_1$ (linear elements) is used to solve the system (50) numerically. Time discretization is performed using an implicit scheme to ensure stability given the stiffness and non-linearity of the problem. The finite element method is preferred over finite difference schemes here, primarily due to its superior ability to accommodate complex boundary conditions. This choice ensures both numerical robustness and accurate enforcement of the physical constraints inherent to the modeled system.

The results of the numerical resolution are presented in the next section for various thin aqueous films containing different initial glycerol fractions.

4.4 Results and discussion

The modeling framework presented in the previous sections has been applied to the numerical study of the isothermal and isobaric drying of thin binary aqueous films composed of water (W) and glycerol (Gl). The liquid is assumed to behave as a Newtonian fluid, with drying conditions that do not promote the formation of a viscoelastic skin at the interface.

Under these assumptions, and for simplicity, the variation of the diffusion coefficient D_W with composition is neglected, setting:

$$D_W(\phi_W) \approx D_W(\phi_W^0) = D_W^0 \chi(\phi_W^0) \tag{51}$$

4.4.1 Mass biot number and drying regimes

In this context, it is useful and possible to define a mass Biot number Bi_m, which compares the internal resistance R_W^d for diffusive transport within the liquid film ($R_W^d = \Delta\phi_W/\Phi_W^{d,m}$) to the resistance R_w^t for external mass transfer in the gas phase ($R_w^t = \Delta\phi_w/\Phi_w^{t,m}$). The Bi_m number can thus be expressed as:

$$\mathrm{Bi}_m = \frac{R_W^d}{R_w^t} = \frac{\Phi_w^{t,m}/\Delta\phi_w}{\Phi_W^{d,m}/\Delta\phi_W} \approx \frac{h_m^{y_w} p_w^{\mathrm{sat}} s}{\rho_W^L p D_W} \tag{52}$$

Note that Bi_m evolves during drying and a small $\mathrm{Bi}_m \ll 1$ indicates that the internal distribution of matter within the film remains approximately homogeneous.

The films studied here share the same physical characteristics as those considered previously in Section 3, except for the presence of an initial nonzero glycerol fraction.

4.4.2 Drying of films with small initial glycerol fractions

We first examine the case of thin films containing small initial volume fractions of glycerol ϕ^0_{Gl} (less than 20%).

In this case, the mass Biot number Bi_m remains below 0.20 throughout drying, suggesting that the assumption of homogeneous concentration fields within the film is acceptable. The matter balances established in Section 4.3 then reduce to the following system of ordinary differential and algebraic equations:

$$\dot{s}(t) = -J(\phi_W(t), t) \quad \text{and} \quad s(t) = s_0 \frac{1-\phi^0_W}{1-\phi_W(t)} \tag{53}$$

where $s_0 = s(0)$ and ϕ^0_W is the initial water volume fraction.

This nonlinear system must be solved numerically.

Figure 11 presents the numerical results for a film with initial thickness $s_0 = 544\,\mu\text{m}$ and initial glycerol volume fraction $\phi^0_{Gl} = 4\%$ (corresponding to $\mathrm{Bi}_m \le 0.18$). Other physical parameters are identical to those used in the pure water film study (Section 4.3).

From **Figure 11a**, it is observed that the model (solid line) captures very well the experimental measurements (symbols), particularly at the beginning and end of drying. Notably, the model accurately reflects the slowdown in drying kinetics due to the increasing glycerol concentration at the end of drying. Furthermore, the solutions (red dashed line) obtained using the simplified system (53) perfectly coincide with the full numerical solution of the PDE system (50), as expected given the low Bi_m values (see **Figure 11b**).

Once fully dried, the film contains glycerol and a residual mass of water $m^{\text{res}}_W \neq 0$ trapped by the glycerol, given by:

$$m^{\text{res}}_W = \rho^L_W \Sigma s_e \phi^e_W \tag{54}$$

Figure 11a reveals an excellent quantitative match between the measured value of s_e and that predicted by the model (dashed horizontal line). This agreement underscores

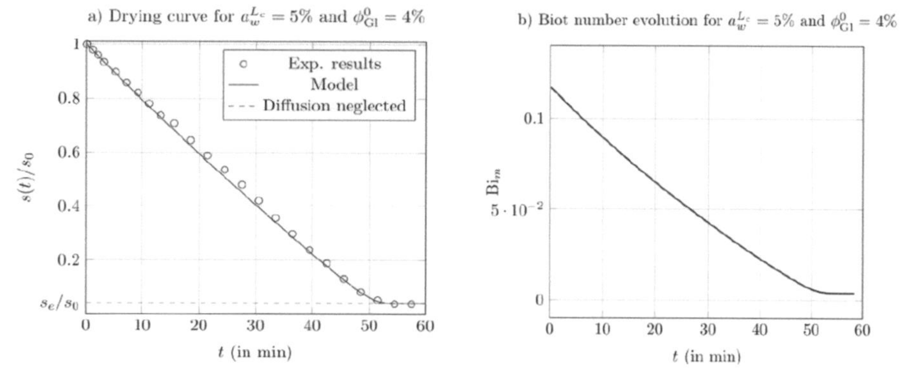

Figure 11.
Isothermal ($T = 298.15$ K) and isobaric ($p = 1$ atm) drying of a thin aqueous film initially containing 4% glycerol ($\phi^0_{Gl} = 4\%$) and $a_w^{L_c} = 0.05$. (a) Temporal evolution of the film thickness; (b) temporal evolution of the mass Biot number Bi_m.

the critical role of water chemical activity a_W in the modeling framework, enabled by the incorporation of a physically sound expression for the activity coefficient γ_W.

4.4.3 Drying of films with high initial glycerol fractions

We now consider films with initial glycerol volume fractions exceeding 20%.

Figure 12 shows the simulation results for a film with $s_0 = 869\,\mu\text{m}$, $\phi^0_{\text{Gl}} = 40\%$, $T = 298.15$ K, $p = 1$ atm, $L_c = 3.45$ mm, and $a_w^{L_c} = 0.05$.

Figure 12a shows the immobilized water content profile $\phi_W(x,t)$ within the liquid film at different times t. For more than half of the drying time ($t_s \approx 140$ min), the profile remains nonuniform, which is consistent with a mass Biot number Bi_m that is not negligible compared to unity (see **Figure 12b**). **Figure 12c** reveals that the mass Biot number becomes negligible only during the final stages of drying, that is, for $t > t_s/2$.

It can be concluded that, unlike in the previous case, it is no longer valid to assume homogeneous composition across the film here, as shown by **Figure 12a** and **b**. **Figure 12c** demonstrates that using the simplified ODE system (53) leads to a poor prediction of the drying kinetics, thus requiring resolution of the full PDE system (50).

4.4.4 Influence of initial glycerol content on drying time and residual water

The drying time t_s is defined as the time required for the film to reach its equilibrium thickness s_e. Predicting t_s as a function of the initial film composition is of primary importance, especially when drying extends over several hours.

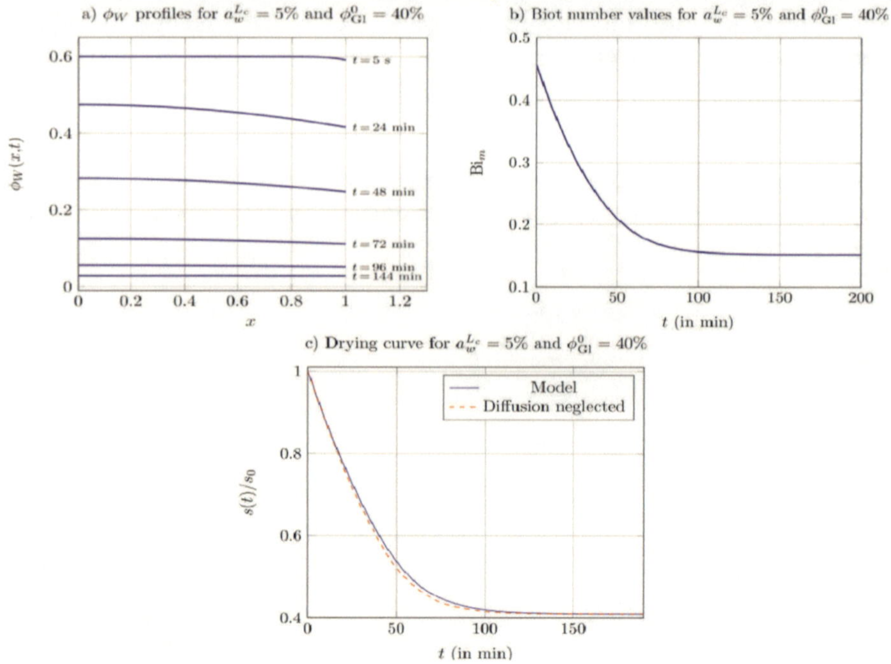

Figure 12.
Isothermal ($T = 298.15$ K) and isobaric ($p = 1$ atm) drying simulation of a thin aqueous film with initial glycerol fraction $\phi^0_{\text{Gl}} = 40\%$ and $a_w^{L_c} = 0.05$. (a) Immobilized profiles of ϕ_W; (b) temporal evolution of Bi_m; (c) temporal evolution of film thickness.

ϕ^0_{Gl} (%)	t_s (min)	m^{res}_W/m^0_W ($\times 10^{-3}$)
1	51	0.25
2.5	55	0.7
5	62	1.45
7.5	68	2.16
10	76	2.97
15	92	4.72
20	112	6.67
25	135	8.92
30	162	11.5
35	194	14.4
40	234	17.8
45	284	21.8
50	351	26.7

Table 1.
Computed drying times t_s and residual water mass fractions m^{res}_W/m^0_W for various initial glycerol volume fractions ϕ^0_{Gl}, with $a^{L_c}_w = 5\%$, $L_c = 3.45$ mm, $p = 101325$ Pa, and $T = 298.15$ K.

Table 1 summarizes the computed drying times and residual water mass fractions for different initial glycerol fractions ϕ^0_{Gl}, considering the same initial mass of water $m^0_W = 29.13$ mg. For each case presented in **Table 1**, the computed value of the ratio s_e/s_0 agreed with the Eq. (49) to within better than 0.1%, indicating strong numerical convergence and mass conservation of the implemented model.

Figure 13 reveals that the drying time t_s increases linearly with ϕ^0_{Gl} for low initial glycerol contents ($\phi^0_{Gl} < 20\%$). Beyond this threshold, a nonlinear increase is observed, reflecting the increasing influence of water-glycerol molecular interactions within the film. A similar behavior is found for the residual water fraction.

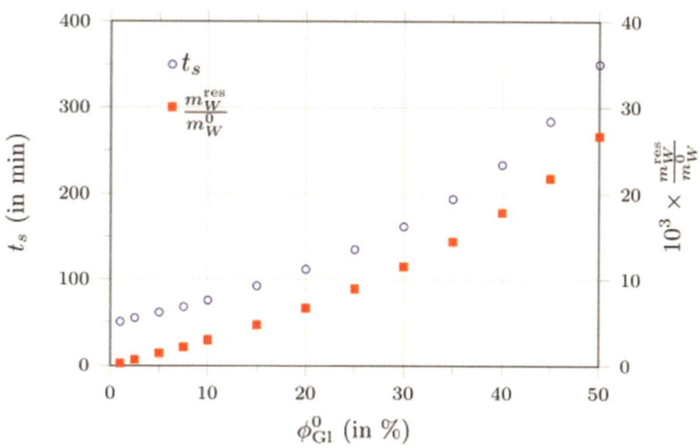

Figure 13.
Evolution of drying time t_s (circles) and residual water fraction m^{res}_W/m^0_W (squares) as a function of initial glycerol volume fraction ϕ^0_{Gl}, for $a^{L_c}_w = 5\%$, $L_c = 3.45$ mm, $p = 101325$ Pa, and $T = 298.15$ K.

5. Conclusion

This work establishes a systemic framework for analyzing the isothermal and isobaric drying of thin aqueous films in the absence of a convective boundary layer. By combining experimental observations, theoretical modeling, and numerical simulations, we have demonstrated that mass diffusion, interfacial dynamics, thermodynamics *via* water activity, and phase change phenomena collectively govern the drying kinetics and the final architecture of the film.

A key finding of this study is the pivotal role of water activity in the drying process. By directly determining the vapor pressure at the liquid-gas interface, water activity controls the intensity of the evaporative flux, the dynamics of the evaporation front, the overall drying time, and the amount of residual water trapped within the dried film. Its gradual evolution—driven by solute enrichment during drying—emerges as a critical lever for predicting and tuning the behavior of complex aqueous films.

The water-glycerol mixture, used here as a model system, proved particularly well-suited for exploring these mechanisms. Owing to its well-documented physicochemical properties and its ability to reproduce behaviors typical of industrial formulations, this binary system enabled the validation of the proposed approach while maintaining a generality that can be extended to other multicomponent aqueous systems of practical relevance.

The model developed in this work accurately captures the coupled dynamics of evaporation and composition evolution, providing a robust predictive tool for the design and optimization of drying processes across diverse application fields such as functional coatings, cosmetics, pharmaceuticals, and biomaterials.

Future research directions include extending the present framework to more complex systems—such as aqueous solutions of water-soluble polymers, sugars, or salts—some of which may exhibit additional phenomena such as viscoelastic relaxation, phase separation, or mechanical instabilities, with the aim of developing a unified and predictive understanding of complex film drying under real-world conditions.

Appendix

The numerical resolution of Eq. (8) is performed using an implicit finite difference scheme with a uniform spatial mesh, defined as $x_i = i\delta x$, for $i = 1, 2, \cdots, N_x - 1$, where $\delta x = 1/N_x$. Time is discretized as $t_n = n\delta t$, with $n = 0, 1, \cdots, N_t$.

The notations

$$v_i^n \equiv v_w(x_i, t_n), \quad s^n \equiv s(t_n), \quad \dot{s}^n \equiv \dot{s}(t_n) \tag{A1}$$

are used for brevity.

The solution column vector at time t_{n+1}, denoted as

$$\left[v^{n+1}\right] = \left[v_1^{n+1}, v_2^{n+1}, \cdots, v_{N_x-2}^{n+1}, v_{N_x-1}^{n+1}\right]^t, \tag{A2}$$

is expressed from the known vector $[v^n]$ at time t_n through the following matrix form:

$$\left[v^{n+1}\right] = \left(\mathrm{ABC}^n\right)^{-1} \cdot \left\{(L - s^n)^2 [v^n] - \left[\mathrm{CL}^{n+1}\right]\right\} \tag{A3}$$

where $\left[\mathrm{CL}^{n+1}\right]$ is the column vector incorporating Dirichlet boundary conditions at $x = 0$ and $x = 1$:

$$\left[\mathrm{CL}^{n+1}\right] = \left[A_1^n v_0^{n+1}, 0, \cdots, 0, C_{N_x-1}^n v_{N_x}^{n+1}\right]^t, \tag{A4}$$

and (ABC^n) is a tridiagonal square matrix of size $(N_x - 1) \times (N_x - 1)$ given by:

$$(\mathrm{ABC}^n) = \begin{pmatrix} B_1^n & C_1^n & 0 & \cdots & 0 & 0 \\ A_2^n & B_2^n & C_2^n & \ddots & \vdots & \vdots \\ 0 & A_3^n & B_3^n & \ddots & 0 & 0 \\ \vdots & \ddots & \ddots & \ddots & \ddots & 0 \\ 0 & \cdots & 0 & A_{N_x-2}^n & B_{N_x-2}^n & C_{N_x-2}^n \\ 0 & \cdots & 0 & 0 & A_{N_x-1}^n & B_{N_x-1}^n \end{pmatrix}. \tag{A5}$$

The matrix coefficients are defined as:

$$\begin{aligned} A_i^n &= -\alpha D_0 \left(1 - \frac{\xi_i^n \delta x}{2D_0}\right), \\ B_i^n &= 2\alpha D_0 \left[1 + \frac{(L - s^n)^2}{2\alpha D_0}\right], \\ C_i^n &= -\alpha D_0 \left(1 + \frac{\xi_i^n \delta x}{2D_0}\right), \end{aligned} \tag{A6}$$

with

$$\alpha = \frac{\delta t}{(\delta x)^2}, \quad \xi_i^n = \left(1 + \beta_w^L - i\delta x\right)(L - s^n)\dot{s}^n, \tag{A7}$$

and the dimensionless parameter

$$\beta_w^L = \frac{\rho_w^L RT}{pM_w}. \tag{A8}$$

The discretization of the jump condition together with the time evolution relation $s^n = \dot{s}^n \delta t + s^{n-1}$ leads to an analytical expression for the film thickness at t_n:

$$s^n = \frac{1}{2}\left[L_c + s^{n-1} - \sqrt{(L_c + s^{n-1})^2 + 4\left(\frac{D_0 \delta t}{\beta_w^L \delta x} \frac{v_0^n - v_1^n}{1 - y_w^s} - L_c s^{n-1}\right)}\right]. \tag{A9}$$

A Python code developed by the authors is available upon request to numerically solve the algebraic system (A3).

Abbreviations

A^0	initial or reference value of A
a_i	activity of species i in a mixture

a_i^*	activity of pure species i
a_w^z	water activity in humid air, at a specific position z
Bi_m	mass Biot number
c	molar concentration of the humid air (mol/m^3)
da	dry air
D	mutual diffusion coefficient (m^2/s)
D_0	molecular diffusion coefficient of water vapor in air (m^2/s)
D_W	diffusion coefficient of water in aqueous solutions (m^2/s)
e_0	initial film width (m)
f_W,f_w	fugacities of water in the liquid and gaseous phases, respectively
g^E	molar excess free energy (J/mol)
h_m^X	mass transfer coefficient relative to X ($\mathrm{kg} \cdot \mathrm{s}^{-1} \cdot \mathrm{m}^{-2} \cdot \mathrm{X}^{-1}$)
$i(t)$, $i_J(t)$	electric current intensities (A)
J	evaporation rate (m/s)
L_c	height of the drying chamber (m)
M_w,M_W	molar mass of water (kg/mol)
$m_W(t)$	mass of water in the liquid film at time t (mg)
m_W^{res}	residual mass of water in the dry film (mg)
$\dot{m}_W(t)$	mass flow rate of water (kg/s)
p	pressure (Pa)
p_e	drying rate (m/s)
p_w	partial pressure of water (Pa)
p_w^{sat}	saturated water vapor pressure (Pa)
R	ideal gas constant ($\mathrm{J} \cdot \mathrm{mol}^{-1} \cdot \mathrm{K}^{-1}$)
RH	relative humidity (%)
RH^∞	relative humidity far from the liquid film (%)
R_W^d	internal resistance for water diffusive transport within liquid films (s/kg)
R_w^t	resistance for external water transport in humid air (s/kg)
$s(t)$,s_0	liquid/gas interface position and initial value (m)
t, t_s	time and drying time (s)
T	temperature (K)
T^{sup}	temperature of the film support (K)
u^∞	air velocity outside the boundary layer (m/s)
v^*	average molar velocity of the gas phase (m/s)
V_i	volume of species i (m^3)
W,W	water in the liquid state
w,w	water in the vapor state
x_i,y_i	mole fraction of species i in a liquid mixture and in a gaseous mixture, respectively
x_i^z,y_i^z	mole fraction of species i at a specific position z, in a liquid mixture and in a gaseous mixture, respectively
$y_w^{L_c,\infty}$	mole fraction of water vapor when $L_c \to \infty$
δ_w	mass transfer boundary layer under forced convection
γ_i	activity coefficient of species i in a mixture
$\Phi_w^{*,t}$	total advective-diffusive molar flux of water leaving the liquid film (mol/s)
$\Phi_w^{t,m}$	total advective-diffusive mass flux of water vapor in the gas phase (kg/s)

$\Phi_W^{d,m}$	diffusive mass flux of water within the liquid film (kg/s)
ϕ_W	volume fraction of water in the liquid mixture
ϕ_W^0, ϕ_W^e	initial and equilibrium volume fractions of water, respectively
$\rho_W^L, \rho_{Gl}, \rho_L$	densities of liquid water, liquid glycerol and liquid mixture, respectively (kg/m^3)
ρ_w^{sat}	density of saturated water vapor (kg/m^3)
τ_0	diffusion time (s)

Author details

Rodolphe Heyd[1*†], Julie Fichot[2], Driss Lahboub[3†], Abderrahim Bakak[3†], Christophe Josserand[4†] and Marie-Louise Saboungi[5†]

1 ENSAM ParisTech, Angers, France

2 Perrigo R&D, Châtillon, France

3 Cadi Ayyad University, Marrakech, Morocco

4 CNRS & Ecole Polytechnique, IP Paris Palaiseau, France

5 Sorbonne University, Paris, France

*Address all correspondence to: rodolphe.heyd@ensam.eu

† These authors contributed equally.

References

[1] Bacchin P, Brutin D, Davaille A, Di Giuseppe E, Chen XD, Gergianakis I, et al. Drying colloidal systems: Laboratory models for a wide range of applications. The European Physical Journal E. 2018;**41**(8):94-128

[2] Fichot J, Heyd R, Josserand C, Chourpa I, Gombart E, Tranchant J-F, et al. Patterned surfaces in the drying of films composed of water, polymer and alcohol. Physical Review E. 2012; **86**(6):1-6

[3] Sekine Y, Ikeda-Fukazawa T. Structural changes of water in a hydrogel during dehydration. The Journal of Chemical Physics. 2009;**130**(3):34501-1-34501-7

[4] Cevc G, Kornyshev A. Simple model of dehydration transitions in lamellar systems, such as lipid membranes. The Journal of Chemical Physics. 1993;**98**(7): 5701-5712

[5] Hopkins RJ, Reid RJ. Evaporation of ethanol/water droplets: Examining the temporal evolution of droplet size, composition and temperature. Journal of Physical Chemistry A. 2005;**109**(35): 7923-7931

[6] Babin V, Holyst R. Evaporation of a sub-micrometer droplet. Journal of Physical Chemistry B. 2005;**109**(22): 11367-11372

[7] Holyst R, Litniewski M, Jakubczyk D, Kolwas K, Kolwas M, Kowalski K, et al. Evaporation of freely suspended single droplets: Experimental, theoretical and computational simulations. Reports on Progress in Physics. 2013;**76**:034601-1-034601-19

[8] Heyd R, Lelong G, Fichot J, Josserand C, Chourpa I, Gombart E, et al. Do we understand the dehydration process. In: Frontiers in Water Biophysics, Perugia, Italy. 2012. pp. 81-82

[9] Heyd R. Biophysics of water evaporation: Theory and experiments. In: MEDICTA 2013 - 11th Mediterranean Conference on Calorimetry and Thermal Analysis. 2013

[10] Fichot J, Heyd R, Saboungi M-L, Josserand C, Gombard E, Tranchant J-F. Drying of polymer films: Study of demixing phenomena. APS March Meeting. 2011;**56**(1):1138. Available from: http://meetings.aps.org/link/BAPS.2011.MAR.Z9.10

[11] Lelong G, Heyd R, Charalambopoulou G, Steriotis T, Brandt A, Beck K, et al. Role of glucose in enhancing stability of aqueous silica gels against dehydration. The Journal of Physical Chemistry C. 2012;**116**(17): 9481-9486

[12] Heyd R, Rampino A, Bellich B, Elisei E, Cesaro A, Saboungi M-L. Isothermal dehydration of thin films of water and sugar solutions. The Journal of Chemical Physics. 2014;**140**:124701-1-124701-8

[13] Bellich B, Elisei E, Heyd R, Saboungi M-L, Cesàro A. Isothermal dehydration of thin films: Calorimetric assessment of model parameters. Journal of Thermal Analysis and Calorimetry. 2015;**121**(3): 963-973

[14] Fichot J. Propriétés de séchage de textures complexes appliquées à la cosmétologie [thesis]. Orléans: Orléans University; 2013

[15] Bird RB, Stewart WE, Lightfoot EN. Transport Phenomena. 2nd ed. New York: Wiley; 2002. p. 895

[16] ASHRAE. Handbook, Fundamentals. SI ed. Peachtree Corners, GA: Technology Parkway; 2021. p. 180

[17] Marrero TR, Mason EA. Gaseous diffusion coefficients. Journal of Physical and Chemical Reference Data. 1972;**1**(1): 1-118

[18] Hennessy MG, Ferretti GL, Cabral JT, Matar OK. A minimal model for solvent evaporation and absorption in thin films. Journal of Colloid and Interface Science. 2017;**488**:61-71. DOI: 10.1016/j.jcis.2016.10.074

[19] Volk A, Kähler CJ. Density model for aqueous glycerol solutions. Experiments in Fluids. 2018;**59**(5):75-1-75-4. DOI: 10.1007/s0034 8-018-2527-y

[20] Sandler SI. Chemical, Biochemical, and Engineering Thermodynamics. 5th ed. New York: John Wiley & Sons, Inc.; 2016. p. 1007

[21] Prausnitz JM, Lichtenthaler RN, de Azevedo EG. Molecular Thermodynamics of Fluid-Phase Equilibria. 3rd ed. Upper Saddle River, NJ: Prentice Hall PTR; 1999. 860 p

[22] Renon H, Prausnitz JM. Local compositions in thermodynamic excess functions for liquid mixtures. AIChE Journal. 1968;**14**(1):135-144. DOI: 10.1002/aic.690140124

[23] Abrams DS, Prausnitz JM. Statistical thermodynamics of liquid mixtures: A new expression for the excess Gibbs energy of partly or completely miscible systems. AIChE Journal. 1975;**21**(1): 116-128. DOI: 10.1002/aic. 690210115

[24] Fredenslund A, Jones RL, Prausnitz JM. Group-contribution estimation of activity coefficients in nonideal liquid mixtures. AIChE Journal. 1975;**21**(6):1086-1099. DOI: 10.1002/ aic.690210607

[25] Gmehling J, Rasmussen P, Fredenslund A. Vapor-liquid equilibriums by UNIFAC group contribution. Revision and extension. 2. Industrial & Engineering Chemistry Process Design and Development. 1982; **21**(1):118-127. DOI: 10.1021/ i200016a021

[26] Zaoui-Djelloul-Daouadji M, Negadi A, Mokbel I, Negadi L. (Vapor-liquid) equilibria and excess Gibbs free energy functions of (ethanol+glycerol), or (water+glycerol) binary mixtures at several temperatures. The Journal of Chemical Thermodynamics. 2014;**69**: 165-171. DOI: 10.1016/j.jct.2013.09.046

[27] Song D, Seibert AF, Rochelle GT. Effect of liquid viscosity on the liquid phase mass transfer coefficient of packing. Energy Procedia. 2014;**63**: 1268-1286. DOI: 10.1016/j. egypro.2014.11.136

[28] D'Errico G, Ortona O, Capuano F, Vitagliano V. Diffusion coefficients for the binary system glycerol + water at 25° C. A velocity correlation study. Journal of Chemical and Engineering Data. 2004;**49**:1665-1670

[29] Okuzono T, Kobayashi M, Doi M. Final shape of a drying thin film. Physical Review E. 2009;**80**(2): 021603-1-021603-11. DOI: 10.1103/ PhysRevE.80.021603